The Admiral's Playbook

The Winning Tactics of History's 15 Greatest Naval Commanders

BILL BODRI

Top Shape Publishing LLC
1135 Terminal Way Suite 209
Reno, NV 89502

Second Edition – July 2026

ISBN-13: 978-1-7370320-8-3

DEDICATION

On behalf of the U.S. Navy, students of maritime history, and the individuals responsible for charting the future course of American naval development.

CONTENTS

ACKNOWLEDGMENTS

A debt of gratitude is extended to Themistocles, and many other maritime greats, whose sea battles made possible the preservation of western civilization. I wish to also thank both Jeffrey Sachs and Martin Armstrong for speaking out to reveal their research and personal opinions on the Russian-Ukrainian conflict in the hopes of injecting some sanity into discussions that might prevent another World War.

CHAPTER 1: THE EVOLUTION OF STRATEGIC NAVAL WARFARE

Before we enter into a discussion of the fifteen greatest naval commanders of all time and their methods, which is a list derived by using multiple AI engines to poll various naval historians, strategists, and famous officers for their opinions on rankings and then finding the commonalities, we need to review a little about the history of naval warfare.

For over two millennia, naval warfare has shaped the destinies of nations and empires, dictating the flow of global commerce. From the ramming *triremes* of ancient Rome to the nuclear-powered aircraft carriers of the United States, the strategies and philosophies of naval combat have evolved in lockstep with changes in technology, geopolitics, and economic needs. Over the last one-hundred years, British and U.S. navies have in particular transitioned from securing trade routes to policing global seas, especially insuring the safety of petroleum trade, and we must briefly revisit the past to see how it has become this way.

Ancient and Classical Naval Warfare (Pre-1500s)

In antiquity, naval battles were mostly an extension of land campaigns, and navy fleets were used to carry troops and secure supply-lines. This simple relationship determined the rise and fall of empires.

In the early eras, the philosophy for using navies was quite straightforward: naval strategy focused on controlling sea lanes to sustain armies and to project power. As a general principle, naval dominance ensured a nation's survival, which was exemplified by Athens during the Peloponnesian War (431–404 BC). Thucydides noted Athens' power came from its maritime mobility to move freely by sea while denying that same freedom to rivals – a principle that would endure through the centuries in various forms.

The ancient maritime empires were economic and political governing systems that were centered around sea routes rather than land territories. Those seafaring states operated by following a fundamental truth, which is to control the waterways to control trade, wealth and overall prosperity.

Whether by land or sea, control over trade routes always translates directly into economic and political power, so maritime states maintained naval forces for the dual purposes of protecting merchant vessels that were the heart of commerce, and projecting influence across vast distances. This strategic priority spurred the development of innovative naval technologies and tactical innovations far beyond those needed for mere transportation.

At this time, naval warfare revolved around just two fundamental strategies – ramming and boarding. Ramming tactics – perfected by vessels like the fast *trireme* armed with a vicious bronze-tipped ram that would act like a guided missile to slice through enemy hulls below the water line – relied on speed, maneuverability, and extremely well-trained crews to deliver devastating attacks. You might maneuver to one side of an enemy vessel and ram it full in the beam (a *periplus* maneuver), or head for a gap between two enemy vessels and then swerve to one side, intending to break off the oars down its length (the *diekplus* maneuver) which would injure all the rowers and leave the ship vulnerable to a beam ram from another vessel.

The Athenians became so proficient at these maneuvers that opponents would flee at hearing of their approach. In contrast, boarding tactics transformed naval battles into infantry engagements, with devices like the Roman *corvus* (a spiked boarding bridge) binding vessels together to allow marines to board ships and fight in bloody hand-to-hand combat. The *corvus*, positioned on a ship's bow, would entrap any enemy vessel attempting to withdraw from a ramming contact.

The choice between ramming and boarding reflected a weighing of both costs and strategic considerations. Ramming, while requiring significant investment in specialized vessels and expert crews, almost guaranteed that targeted ship would be disabled and sunk, thus delivering quick victories without the heavy casualties of prolonged battles. However, it carried the risk of entangling the ram, which could result in the attacker being stuck, captured, or even sunk by their own collision if their hull was compromised.

Boarding, conversely, capitalized on existing military skills and preserved valuable enemy ships as potential prizes, but at the cost of brutal hand-to-hand close-quarters combat. Nevertheless, a captured ship could be added to one's own fleet and the cargo and crew of a captured ship could be sold for massive profit. Ramming eliminated all these potential profits by sinking the prize.

Naval power at this time demanded enormous resource investments because a single *trireme* not only necessitated ship building costs but required funding 170 skilled oarsmen plus extensive logistical support. These economic realities ultimately determined which tactical approach dominated as time passed.

Because of the costs, medieval naval warfare therefore shifted toward boarding-based tactics that would thus dominate until the advent of effective artillery. Ramming tactics vanished by the 7th century AD as ships grew sturdier and economies could not support the costs of specialized vessels and crews anymore. Medieval naval combat then pivoted to a new dominant approach, which was to board an enemy's ship with sailors trading ship-to-ship blows rather than vessel-to-vessel crashes. This hand-to-hand combat paradigm would rule the waves until effective artillery (cannons) emerged much later.

Ancient naval warfare definitively established several principles for national supremacy, most importantly being the fact that maritime control enabled trade dominance, power projection, and imperial expansion. Ingenious innovations like the *corvus* boarding bridge and Greek fire, along with sophisticated fleet tactics and evolving ship designs, marked some of mankind's earliest attempts to systematically master ocean warfare, and created naval principles that still guide naval commanders and strategists centuries later.

We can also, in a more detailed fashion, chart the evolution of naval strategies over time in a more useful way by grouping naval warfare into several logical phases. This starts with the "Age of Sail."

The Age of Sail (1500s–1800s)

The Age of Sail was a ~300-year era in which long-distance ocean travel, global trade, naval warfare, exploration, and empire-building were overwhelmingly powered by sailing ships. It marked the first time in history that humans had mastered ocean travel enough that they could reliably cross and dominate all the world's oceans. Essentially, sailing ships enabled truly global trade and travel for the first time. They were the primary means of long-distance transport for goods, people, and military power until the introduction of steam-powered ships and the launching of ironclad warships in the mid-19th century.

Spanning the 16th to 19th centuries (roughly 1570 to 1860), the Age of Sail made naval warfare the essential center of economic prosperity and imperial power. This historical era was all about wealth and power achieved via the sea. Global trade networks – Spanish silver fleets, Dutch East India Company (VOC), British East India Company, Atlantic triangular trade (slaves-sugar-rum), Manila galleons, tea/silk/porcelain routes and so forth made nations rich if they had ships to engage in international trade. The principle nations followed was that command of the sea secured colonial riches and produced geopolitical supremacy.

This era's geopolitical philosophy hinged on controlling sea lanes (trade routes) to fuel empires, with prosperity, wealth and power flowing to those who mastered the oceans. Robust, fully rigged sailing ships capable of carrying heavy naval artillery made this possible. Decisive naval battles between the great fleets of nations like Great Britain, France, and Spain fought battles (Spanish Armada 1588, Trafalgar 1805, etc.) that decided who controlled colonies, trade routes, and ultimately global power. These battles decisively demonstrated how naval supremacy dictated the fate of nations, proving that sea control equated to global dominance and wealth.

The life for sailors was famously difficult characterized by cramped conditions, poor rations, harsh discipline (flogging), disease (especially scurvy due to lack of vitamin C), constant dangers from the elements, and press gangs. Ships became larger and more complex over this period, evolving from ships of exploration to the powerful frigates, ships-of-the-line, and later, the speedy clipper ships that marked the period's final flourish (the "Golden Age of Sail," 1840s-1860s).

Naval strategy in this era centered on concentrated ship firepower, disciplined fleet formations, and ship maneuverability. Full-rigged ships, frigates, galleons, East Indiamen, clippers, and warships all traversed the waves over this era but ships-of-the-line, heavily armed with up to 120 cannons, eventually formed the backbone of power projection, using line tactics to maximize artillery broadsides against enemy vessels. In the line of battle tactic, man-of-war ships formed a column to maximize the use of their broadside cannon fire. The naval philosophy for dominance consisted of controlling the seas by securing vital trade routes, protecting merchant fleets, and denying resources to enemies through blockades.

Naval supremacy wasn't just about winning battles because it required a network of strategic bases like Gibraltar and Jamaica that enabled empires to project overwhelming force worldwide. Port cities therefore served as power projection hubs across the globe. The Dutch, Spanish, French, and British all understood these fundamental principles, which is why they fought so fiercely for maritime dominance as the ultimate path to global dominance and riches. Over time the major power dominance shifted from Portugal to Spain to the Dutch Republic and then Britain (which went unchallenged from 1805–1860s).

This rivalry between nations ignited a remarkable era of innovation that transformed ship designs, revolutionized ship gunnery, and improved navigation techniques as each power sought the upper hand. The naval powers understood that naval innovations would determine which empires would rise or fall.

While grand fleets clashed for supremacy during this age, privateers and commerce raiders didn't merely supplement fleet actions. They acted to strategically cripple enemy economies, as dramatically demonstrated in the Anglo-Dutch Wars (1652-1784). This relentless competition for control of the seas defined the Age of Sail and created the foundation of naval warfare strategy we recognize today. It literally shaped the international power dynamics of our modern global order.

The Age of Sail created the first truly global economy, and decided which nations would become superpowers. It ended because of innovation when steam and steel took over in the 1860s.

The Industrial Age and Mahanian Doctrine (Late 1800s–Early 1900s)

The subsequent era of the Industrial Revolution revolutionized naval warfare with the invention of steam-powered ironclads, rifled guns, and torpedoes. Wooden hulls transformed into iron and steel armor, and the marine steam engine replaced the sail so that vessels were no longer dependent on the wind but could maintain consistent speed and schedules. They could take shorter, less wind-favorable routes. By the 1870s, sailing ships were quickly relegated to carrying bulk, low-value cargo on long voyages where time was not critical, and by the 1920s and 30s, they had largely disappeared from commercial trade, marking the definitive end of the era.

Alfred Thayer Mahan's *The Influence of Sea Power Upon History* (1890) brilliantly articulated what would become the defining strategic principle of the era, which is that national greatness, prosperity and military supremacy fundamentally depends on control of the sea (naval supremacy). He asserted that a powerful navy must aim not merely to defend its coast, but to dominate the major sea lanes of the world.

Mahan believed that naval supremacy comes only through concentrated battle fleets capable of destroying enemy navies in decisive engagements. His core thesis, in one sentence, is that "Whoever controls the sea controls trade, wealth, and ultimately world power; history proves that great nations become great by building large battle fleets, overseas bases, and fighting for 'command of the sea'."

Mahan's doctrine proved so compelling that major powers immediately restructured their naval strategies around controlling crucial sea lanes and strategic chokepoints and establishing naval bases (like Great Britain's empire). His emphasis on maritime strength for controlling sea lanes and chokepoints inspired the U.S. "Great White Fleet," the German High Seas Fleet, Japanese expansion, and the pre-WWI naval arms race.

He argued that America should acquire strategic bases across the Pacific and Caribbean (Hawaii, Guam, Puerto Rico, Panama). Essentially, Mahan preached having a big navy, big battles and global baes to secure world power, and made sea power synonymous with global power.

Julian Corbett's *Some Principles of Maritime Strategy* (1911) countered Mahan's viewpoint by advocating limited war and the ingenious "fleet-in-being" concept, which meant using navies to exert economic pressure and deter enemies without risking destruction. Where Mahan saw sea power as an end in itself, Corbett argued that naval strategy is a subdivision of national (grand) strategy and that control of the sea is a means, not an end. Corbett believed that the goal is to secure temporary local control over specific sea lanes only when and where this was required for specific purposes. He argued that you rarely sweep the enemy fleet from the ocean and just need enough control to use the sea while denying it to the enemy when/where it matters (the "fleet in being" concept).

Corbett also demonstrated how navies could project power, strangle enemy economies, and achieve objectives without risking catastrophic fleet losses. His emphasis on joint operations and commerce protection was a pragmatic alternative to Mahan's battle-centric approach and its all-or-nothing battle mentality. Essentially, he shifted naval theory from the pursuit of Mahan's singular, decisive battle to a Clausewitzian strategy of achieving national goals through the calculated use of the sea.

Corbett's ideas on control of trade routes and amphibious operations (rather than the annihilation of the enemy fleet) became the actual British way of war (which was using blockades and expeditionary ops). It influenced the Royal Navy's WWI strategy to avoid risky decisive battles while using distant blockades, and his book remains the foundation of the "British school" of maritime strategy that is still required reading at most Western naval colleges. Its message can be distilled into the thought: "Naval force exists to secure the communications on which national power depends; decisive fleet battles are rare and usually unnecessary if you control the sea lanes and use amphibious power wisely."

Stepan Makarov, the visionary Russian vice-admiral, also revolutionized naval thinking by instead prioritizing technology, tactical flexibility that favored torpedo boats and coastal defenses, and the human element over the pure fleet size or decisive fleet battles. His writings on naval tactics noted that "a ship is a fool to fight a fort," anticipating later debates about coastal defenses versus Mahan's massive fleet power. He was the world's leading advocate for the self-propelled torpedo and destroyer, stating, "A few bold torpedo boats at night can cripple a battleship squadron and change the whole strategic situation."

As a proponent of aggressive limited-war naval tactics, the torpedo revolution, and initiative, Makarov's ideas were 20–30 years ahead of most admirals of his time. He essentially sketched the blueprint for modern "maneuver warfare at sea" (asymmetric warfare) where small, fast, homogeneous squadrons that can concentrate quickly beat large, slow, mixed fleets that arrive piecemeal. His ideas left an indelible mark on shaping the approaches of smaller navies worldwide, basically providing them with a blueprint to maximize their limited resources against stronger opponents just as drone warfare is doing today.

Basically, the pursuit of naval supremacy defined this era's warfare thinking. Everyone built dreadnought battle fleets due to the influence of Mahan telling what to build, Corbett's strategies told Britain how to win WWI, and Makarov's ideas showed how the future would evolve with developments in torpedoes, mines, aggressive small-unit tactics and later air-power. Whether through Mahan's decisive battles, Corbett's strategic deterrence, or Makarov's technical innovations, powerful nations transformed their fleets into tools for expanding their empires and controlling global commerce.

The World Wars and the Carrier Revolution (1914–1945)

The era of the World Wars shifted naval strategy from the centuries old "gun vs. gun" to "aircraft vs ship." Naval warfare shifted from battleships to carriers, evident in Japan's 1941 Pearl Harbor attack where six Japanese fleet carriers destroyed or damaged eight American battleships. This attack exposed battleships' vulnerabilities to air strikes that put an end to various naval doctrines. The Age of Sail had been replaced by the Age of Steam, and now the Age of the Battleship was permanently replaced by the Age of the Carrier.

The decisive realization was that a warships' striking range was no longer limited by the distance its guns could fire, but by the distance its aircraft could fly. Carrier aircraft now gave range (300–500 km vs battleship guns ~35 km) and first-strike capability to navies. In terms of design, aircraft advanced to carry meaningful payloads such as torpedoes and heavy bombs. A single well-timed air strike could now sink or cripple capital ships before they ever came into gun range. Battleships therefore became floating anti-aircraft batteries whose main job was to protect the carriers and support amphibious landings, not to fight other battleships even though this had been the pattern or centuries.

The evidence that carriers had become the significant ships became undeniable at Midway (1942) where Raymond Spruance's carrier tactics achieved what was previously unthinkable: the destruction of four Japanese carriers through superior intelligence, strategic deception, and air strikes that sunk the Japanese ships without the main surface fleets firing a single shot at one another. The victor was determined by whomever could launch the first, best-coordinated airstrike. Spruance's philosophy of prioritizing information, timing, and calculated risk won a battle and revolutionized naval warfare for generations to come.

The battles of the Coral Sea (1942) and Midway proved that fleets could engage without ever seeing one another, another revolution in naval warfare. In short, carriers became the linchpin of sea power, redefining the ideas of naval supremacy that had previously focused on battleships. It could find enemies hundreds of miles away, attack many targets simultaneously, strike an enemy before it could strike you, and carry sufficient aircraft to replace the firepower of an entire battle line. The new philosophy valued air power, flexibility and reach over raw tonnage, where the carrier became the new capital ship.

Naval warfare during WWII also pivoted to economic targets because navies could choke trade routes and thus become economic weapons. Germany's Erich Raeder understood that Germany could not match Allied naval production, so he adopted asymmetric strategies to disrupt sea lanes rather than control them. Raeder favored commerce raiding, deploying U-boats (as auxiliaries to surface fleets) against Allied supply lines to isolate Britain, which was a strategy that nearly succeeded in starving Britain into submission. His successor, Karl Dönitz, used U-boats as the primary instrument in the attempt to strangle Britain. The Allies countered by developing anti-submarine warfare (ASW) tactics like convoy systems and harnessing sonar for submarine detection to hunt the submarines.

Allied control of the Atlantic, secured through escorts and air cover, helped secure WWII victory in Europe by safeguarding critical trade routes. Allied navies also slowly strangled Axis logistics through relentless blockades. The carrier revolution embodied Julian Corbett's joint operations theory, and established sea power as a blend of air, surface, and subsurface forces, which is the foundation of modern doctrines of networked warfare.

The Cold War and Nuclear Deterrence (1945–1991)

The Cold War led to a new era of naval evolution dominated by the introduction of nuclear weapons and the systemic geopolitical rivalry between the United States and the Soviet Union. One surprising development was the shift in focus from traditional surface engagements to undersea warfare. The Cold War transformed navies into instruments of power projection, global deterrence, and geopolitical influence through the innovations of nuclear submarines, ballistic missiles, and satellite reconnaissance that, together with carriers as the new capital ship, replaced the battleship as the new measure of sea power.

Soviet Admiral Sergei Gorshkov provided the intellectual foundation for the Soviet Navy's transformation from a coastal-defense force into a true blue-water, global power. Gorshkov envisioned the navy as a political instrument, globally present, and balanced force (not battleship-centric) that protected state interest, supported diplomacy in peacetime while still being able to fight a nuclear war at sea. His balanced fleet deployed surface ships (missile cruisers), submarines, naval aviation and marines to counter U.S. carrier dominance while projecting Soviet influence. The Soviets focused heavily on developing weapons to take out U.S. carrier battle groups using long-range anti-ship missiles launched from these multiple platforms. Therefore Gorshkov's balanced fleet concept – surface ships (missile cruisers), submarines, naval aviation and marines – challenged Western sea control through a persistent naval presence worldwide.

U.S. carrier battle groups countered by providing a forward presence that deterred Soviet expansion. They supported conflicts like Vietnam, and essentially embodied Mahan's view of sea power as national strength.

Naval philosophy at this time transitioned from destruction to deterrence, with nuclear submarines ensuring mutual assured destruction because of their warhead missiles. A single nuclear strike could potentially wipe out a concentrated fleet so this led to a strategy of dispersal and the need for greater maneuverability.

Rather than decisive battles, J.C. Wylie's *Military Strategy: A General Theory of Power Control* (1967) produced a blueprint for Cold War naval operations that promoted cumulative attrition through dispersed persistent operations. In other words, he advocated a cumulative strategy of many small, persistent, dispersed operations that added up rather than a single decisive battle to establish and maintain control over the enemy's choices and actions. His focus on control over destruction encapsulated the idea that Mahan's approach was suicidal, and that victory came from wearing down the enemy over the years via a "long twilight struggle" at sea instead of a Jutland-style showdown.

Naval officer and CIA director Stansfield Turner effectively killed the last remnants of the Mahanian "big-fleet battle" mindset in U.S. naval strategy. He understood that precision weapon strikes (against a handful of targets that actually mattered) guided by accurate intelligence could accomplish what massive fleet engagements could not, so he championed a tech-driven warfare of precision strikes as surgical operations instead of fleet battles.

Turner saw that satellites, signals intelligence, and soon stealth plus cruise missiles were making the classic carrier-vs-carrier duel obsolete, and that it was foolish to risk thousands of sailors' lives and dozens of aircraft when three Tomahawk missiles guided by a satellite could sink a Soviet carrier from standoff range hundreds of miles away. Instead of building ever-larger carrier groups to slug it out with Soviet surface forces, he therefore refocused the U.S. Navy strategy on using real-time intelligence to locate high-value targets that it could then kill from an extreme distance using Tomahawk cruise missiles, Harpoon anti-ship missiles or carrier aircraft guided by AWACS and satellite data.

Naval supremacy now depended on technological advantages in sonar, missiles, and satellites, while its forward presence was used to deter Soviet expansion and accomplish other objectives. By the end of the Cold War, naval strategy had fully integrated the nuclear deterrent role of the SSBN (a nuclear-powered submarine designed specifically to carry and launch nuclear-armed ballistic missiles) and the power projection role of the carrier, while making the eternal search for the enemy submarine the core tactical challenge. The carrier battle group was now seen not just as a defensive shield but as a highly potent offensive weapon capable of striking deep inland against Soviet naval and air bases.

Basically, the Cold War turned navies from battlefleet duelists into permanent global police and nuclear deterrence platforms (SSBNs). The idea became to show the flag, deter war and fight only limited wars with surgical strikes whenever possible. Navies became truly global 24/7 forces with SSBNs always at sea; carriers could shadow enemies at a distance or stay on station off someone's coast indefinitely. Carriers and helicopter ships became floating bases for limited wars and interventions. Navies evolved into instruments of political leverage, economic safeguarding, and crisis response, shaping a American-Soviet world where control of the seas underpinned global stability.

The Post-Cold War to Present (1991–2025)

When the Cold War ended, navies worldwide underwent yet another transformation. Post-Cold War navies shifted from American-Soviet deterrence to great power competition with China and Russia, power projection against non-state actors, and policing and countering asymmetric threats in a multipolar world. This period has seen the continued dominance of the aircraft carrier while heavily integrating network-centric warfare, precision-guided weapons and new developments such as drones, unmanned surface and underwater vehicles, and AI. There is also a priority on developing and deploying of ship-launched hypersonic missiles.

Naval power during this era has also evolved beyond traditional combat to tackle humanitarian missions, expeditionary operations, and secure global waterways, with U.S. carrier groups demonstrating unrivaled power in the Iraq and Afghanistan conflicts.

Going back to 1986, Wayne Hughes' *Fleet Tactics* introduced the naval strategy of "distributed lethality" – using small, missile-armed vessels against larger forces. His basic premise was that we were now in the missile age when victory no longer belonged to the side with the biggest or most ships, but to the side that first detects, first decides, and first shoots the most missiles, which you could achieve with many cheap, dispersed, heavily-armed small units instead of a few golden carriers.

His insight that dispersed forces survive better against numerically superior foes, that the number of shooters was more important than the concentration of hulls, and that missiles decide battles, reshaped naval doctrine. The basic idea was to build a cheap swarm of hundreds of small, stealthy, missile-armed ships and scatter them across the ocean so that the enemy could never find and kill them all (a distributed missile-heavy fleet) before the first salvo.

China's rise since the 1980s has also threatened the established order. Toshi Yoshihara and James Holmes, in their 2010 work *Red Star Over the Pacific*, identified a Chinese A2/AD strategy using advanced precision-guided missiles, quiet submarines, long-range sensors and layered defense systems (instead of battleships) to turn its First and Second Island Chains into no-go zones for U.S. surface forces, thereby achieving sea control inside those chains without ever needing a blue-water battlefleet. A2/AD was a way to make it politically unacceptable for the U.S. to risk multi-billion-dollar carrier groups within 1,000–1,500 nm of the Chinese coast, effectively pushing the U.S. Navy away from its coasts and back to Guam/Hawaii, thus breaking the post-1945 American security guarantee in the Western Pacific.

This represents nothing less than a direct assault on Mahan's sea-control doctrine that has underpinned American naval dominance for generations. The U.S. response has been Hughes' distributed lethality strategy where instead of relying on a few highly capable, high-value assets (like a single aircraft carrier) that are vulnerable to A2/AD weapons, the goal has been to equip a larger number of smaller, less expensive ships with offensive anti-ship missile capabilities.

Andrew Krepinevich and Sam Tangredi (*Maritime Strategy in a New Era*, 2012) next declared that the missile age had permanently replaced the gun/aircraft age and that future naval war would be decided in the first 30–60 minutes by whoever fires the biggest, fastest, most complex missile salvo that the enemy cannot stop. They highlighted "missile salvo warfare" of massive, hyper-velocity missile exchanges in which the side that launches more missiles than its enemy can shoot down will win instantly, so victories will hinge on launching overwhelming first strikes that saturate enemy defenses.

Essentially, the supercarrier-centered navy was mathematically dead in a peer missile fight. The missile salvo itself has become the new capital ship of naval fleets, and the present view is that any navy that cannot fire (or stop) 1,000+ missiles in the first hour will cease to exist as a fighting force in the first day of the next major war.

Krepinevich and Tangredi also argued that navies must now adapt to asymmetric threats – drones, swarms, USVs and UUVs, mines, cyber-attacks, and hypersonic weapons – that level the playing field between great powers and smaller adversaries. The fact is that affordable, mass-produced advanced technology now enables weaker opponents to challenge or even triumph over major naval powers. They can impose catastrophic costs on the most expensive, high-signature platforms (carriers, destroyers, amphibious ships, bases). Their view is that any navy that keeps investing primarily in a few exquisite platforms instead of thousands of cheap, smart, dispersed killers is choosing to lose the next war before it starts. In short, the mindset now is that controlling the seas demands innovation along with technological and strategic agility.

Naval Strategies and Reserve Currencies

Tactics win battles, but economics sustains fleets. Having summarized the tactical evolution of naval warfare, we must now examine it through a different lens: the financial and fiscal angle that underpins maritime power.

Since the current U.S. Navy's global posture and key strategic missions are deeply intertwined with defending the Petrodollar system and the fact that the USD is the world's reserve currency, we will look at the evolving naval strategies of those nations whose currency once functioned as the dominant or co-dominant international reserve, trade, and debt-settlement currency since that is the United States situation today. This financial history provides the essential context for evaluating modern strategies – a task for our final chapter.

There is no historical timeline of currencies having neat, successive dominance. For instance, the Spanish dollar and Dutch guilder circulated simultaneously for decades. Also, the Pound sterling was dominant while the French franc was simultaneously important in specific regions, and the Venetian Ducat and Florentine Florin were dominant in earlier periods than we will examine.

As a simplification, we will particularly examine the naval strategies of Portugal, Spain, Dutch Republic, France, United Kingdom, and United States in turn because the preferred currencies for global trade over the past few hundred years have roughly been as follows:

- 1420s–1530s → Portuguese real
- 1530s–1640s → Spanish piece of eight
- 1640s–1780s → Dutch guilder
- 1720s–1815 → French franc/livre (shared with sterling)
- 1815–1914 → British pound sterling
- 1914–today → U.S. dollar

Portuguese Naval Strategy (1450s–1530s)

Portugal's naval strategy during the Age of Exploration established it as the first global maritime empire. Under the visionary leadership of Prince Henry the Navigator and Vasco da Gama, Portuguese ships ventured where none had dared before, pioneering oceanic routes that revolutionized oceanic navigation forever. The Portuguese Navy was less a traditional fighting force and more the armed envoy of a mercantile state, its cannons backing the ledger books of Portugal. Portugal focused on establishing trade monopolies in spices and gold by securing routes from West Africa to India that were controlled through its naval presence.

The philosophy of Portuguese naval strategy emphasized exploration, the establishment of fortified trading posts *(feitorias)* to sustain long-range operations, and the deployment of small but agile fleets. Violence was kept to a minimum by seizing a coastal fort, leaving behind an fortified trading post *(feitoria)* to control commerce, and sailing away without attempting to seize territory. The Portuguese never sought decisive fleet battle unless the odds were overwhelming, but Diu 1509 was an exception. They preferred to protect their commerce (using warships to escort an annual convoy of spices home) and employed commerce raiding against non-compliant traders.

Portugal had discovered that, in an Age of Sail, a tiny but technologically superior navy, operating from a network of fortified choke points, could extract far more wealth than a large traditional battlefleet ever could – a strategy that kept Lisbon the richest city in Europe for almost a century while spending a remarkably small percentage of GDP on the navy (≈3–5% vs. Spain's 20–25% on its Armada).

Geopolitically, Portugal used its navy to bypass Muslim-controlled land routes, connecting Europe to Asia and Africa. The commercial payoff was a spice trade monopoly that made Lisbon Europe's wealthiest city. The Portuguese real consequently became the first global reserve currency as the preferred coin for Asian and Africa trade although Vencie's ducat co-existed at the time.

Portugal's wealth flowed directly from its naval supremacy, built upon technological superiority (advanced caravels and astrolabes), environmental mastery (monsoon navigation), and strategic outposts like Goa and Malacca. Portugal kept its standing navy tiny (rarely more than 30–40 ocean-going ships worldwide) because the real profit came from the licensing system *(cartaz)* and tax farming at the choke-point fortresses; every ship in the Indian ocean had to buy a Portuguese *cartaz* (naval pass) or be seized as prize. When the trade monopoly began to slip in the 1550s-1570s, Portugal doubled down on an asymmetric low-cost defense by using galleys and light oared vessels in narrow waters (Red Sea, Persian Gulf) and scorched-earth raids (burning enemy ships in harbor) instead of expensive standing squadrons.

The Portuguese empire was essentially the first "Corbettian" maritime power since it used sea control as a primarily economic weapon rather than a military end in itself, and used minimal force to maximize economic gain.

Spanish Naval Strategy (1530s–1640s)

At its imperial peak, Spain's naval strategy focused on defending its vast empire and securing the flow of silver from the Americas into its treasury. Under monarchs like Philip II, the Spanish Navy combined galleons with galleys to protect silver treasure fleets (that funded imperial ambitions) and project power in the Mediterranean and Atlantic. Initially Spain's navy was focused on exploration but then switched emphasis to the convoy system, transitioning from a Mediterranean galley focus to protecting Atlantic galleons that were transporting treasure. However, if we look at the Spanish naval strategy from a prosperity perspective we can see that over time the navy's usage progressively transformed from aggressive wealth-creation to increasingly desperate (and ultimately failed) wealth-preservation.

Spain's navy was the aortic artery of its empire, protecting Mediterranean trade and pumping American silver – the lifeblood of Habsburg power – to the heart of Europe. The philosophy of naval warfare initially blended defensive convoy systems (two escorted fleets per year that enabled 90-95% of silver to reach Seville safely) with offensive campaigns, as seen in the unsuccessful 1588 Armada that struck at England aiming to disrupt its maritime ambitions. In terms of evolution, post-1588 the Spanish navy started avoiding battles because of the Armada experience, next transferred from using Spanish-built ships to hiring Genoese/Flemish shipping because Spanish shipyards were too expensive and slow after 1620s, and then (because the Crown went bankrupt) Spain eventually went from having a silver monopoly to selling the rights to carry silver to licensed foreign Dutch/English/Genovese traders.

Spain had designed an imperfect but effective naval defensive system to protect the greatest revenue stream in history, but the more silver output eventually declined the more its navy had to be used to protect the dwindling remainder. This system reached a point of diminishing returns where the escorting galleons eventually cost more than the silver they protected. When the resource flow collapsed, which is the typical fate of commodity production, the naval protection fleet became the Spanish empire's single biggest money-loser, which has relevance for later lessons in this book.

Geopolitically, while safeguarding its lucrative trade routes from privateers Spain the navy was also used to enforce Catholic dominance over Protestant powers. The trade route payoff of immense wealth from New World silver funded Spain's European supremacy, so this was the focus of attention because the silver ships from the New World were eventually the primary source of Spanish prosperity. Over this time, the Spanish silver dollar (*real de a ocho*, or piece of eight) became the first true world dollar minted in enormous quantities from this silver, and was accepted nearly everywhere because it was the only coin guaranteed to be full-weight silver.

After the Armada disaster, Spain abandoned the "decisive battle" idea. Spain's supremacy then relied on key fortified ports like Havana, Cartagena, Veracruz, Manila and Cádiz together with a two-ocean navy and defensive convoy warfare. However, vulnerabilities to nimble Dutch and English raiders exposed a critical flaw, which was that Spain's reach exceeded its ability to protect its foreign presence. Spain gradually adopted a "silver-first" doctrine where everything became subordinated to getting the fleets home, and by the 1630-40s period nearly 70-80% of the entire Spanish navy was tied up escorting its two treasure fleets per year at a loss, a financial ruin for the Empire instead of the low-cost trade model that had worked for Portugal a century earlier. Spain basically bled itself protecting shrinking silver flows with an ever-larger navy. This story bears relevance to America's practice of policing the world's oceans as long as the world's oil trade uses only the USD.

Dutch Naval Strategy (1640s–1720s)

During the Golden Age, the mercantile Netherlands used its Navy to protect its vast trade empire by creating a sophisticated wealth-preservation system that worked a century longer than the Spanish economic model. The Dutch Navy functioned as the shield for a trading republic, escorting merchant fleets to keep the world's most efficient merchant marine profitable. Under admirals like Michiel de Ruyter, the Dutch navy focused on protecting merchant convoys and controlling key trade routes in the

Baltic and East Indies. Eventually the Dutch were sending out constant patrols to escort ships and hunt privateers.

During the 1620s-1650s, the VOC (*Vereenigde Oostindische Compagnie*), or Dutch East India Company, created the first permanent ocean-going battlefleet. Dutch freight rates at this time were 30-50% lower than even English and French competitors carrying trade. To understand the wealth this entailed, the VOC dividends reportedly averaged around 18% annually for much of the 17th century, peaking at 40-75% in good years.

The Dutch philosophy of naval warfare basically prioritized economic security over territorial conquest. Their shallow-draft vessels excelled in maneuverability for riverine and coastal operations, notably during the 1667 Raid on the Medway where de Ruyter humiliated the English navy, and forced England to sign peace and accept Dutch commercial terms. Dutch convoy losses fell to near zero during the three Anglo-Dutch Wars, and afterwards the Dutch navy shifted to using a combination of convoys and blockades instead of seeking decisive battle.

The Dutch scaled its navy to the size of its trade so when trade volumes fell after 1740, they simply cut the fleet and lived off financial services (bills of exchange, marine insurance, acceptance loans). Eventually, the Dutch Republic outsourced its naval protection to the British Royal Navy in exchange for commercial privileges.

Before it came to this, however, geopolitically the Dutch Navy was always defending against threats from both England and France, serving as the backbone that protected the global network of the Dutch East India Company. The basic strategy was defensive-offensive commerce protection. This protection brought the Dutch unrivaled wealth and prosperity, with Amsterdam replacing Antwerp as the financial center of Europe. The Bank of Amsterdam's Dutch guilder became the most trusted bill of exchange and the de facto reserve currency. By the Fourth Anglo-Dutch War (1780–84), the Netherlands had only 19 ships-of-the-line left and was crushed, but the commercial infrastructure of the Bank of Amsterdam and its commodity markets survived another generation. In fact, the Bank of Amsterdam remained the safest deposit in Europe until the 1780s.

Overall, the Dutch naval doctrine had emphasized convoys and amphibious operations with strongholds like Batavia providing strategic footholds. While the Anglo-Dutch Wars demonstrated tactical innovation with Admiral de Ruyter's line-of-battle formations, by 1720 the golden era had ended. In essence, British ascendancy ended Dutch naval dominance, which was surrendered essentially because the Dutch prioritized commerce over empire.

Even so, the Dutch had created the first truly commercial-maritime strategy: the navy existed solely to keep freight rates low and insurance premiums lower. When that was no longer possible with their own warships, they had quietly switched to British protection while continuing to earn 2–3 % of European GDP from shipping and finance alone until the Napoleonic Wars. In modern terms, the Dutch Republic ran the 17th–18th century equivalent of a Singapore or Hong Kong – a giant import-export center whose "defense policy" was to make itself indispensable to whoever happened to be the reigning naval superpower.

French Naval Strategy (1720s–1815)

France's naval strategy during the 18th and early 19th centuries aimed to challenge British maritime dominance while supporting its own colonial and economic ambitions of a global trade empire. The idea was to cripple a wealthier maritime rival without matching its fleet ship-for-ship. The French Navy, under ministers like Jean-Frédéric Phélypeaux de Maurepas, prioritized a *guerre de course* (commerce raiding) to disrupt British trade and weaken Britain's foundation, coupled with fleet actions to contest sea control. France basically instituted a deliberate "low-cost economic bleeding" strategy against its rival Britain that had a global reach.

France's geopolitical strategy was to undermine Britain's economic lifeline rather than match its fleet. France discovered the most cost-effective naval formula in the 18th century was to spend 60–70% less than Britain on the battlefleet, but use geography, privateering, and selective intervention to make Britain spend 3–4 times more to defend its trade. This worked brilliantly in every war until Trafalgar and the loss of Saint-Domingue, a cash cow that won independence from French control in 1804 (whose sugar and coffee exports, before the slave revolt of 1791, were worth more than all of Britain's 13 North American colonies combined) destroyed the colonial revenue base that paid for its privateers. After 1815, France never again had the fiscal strength to run a true global navy while Britain did, but would keep its empire together with a tiny battlefleet that showed the flag at minimal cost.

The French philosophy for dealing with the British during this era emphasized flexibility, using privateers and fast frigates to harass merchant shipping, and maintaining a battle fleet to deter or engage the Royal Navy when it was advantageous – never fighting the Royal Navy unless it had local superiority. This strategy of selective fleet actions was seen in American Revolutionary War's Battle of the Chesapeake (1781), where Admiral de Grasse's victory secured Yorktown's surrender with only 60% of the British battlefleet present.

The French *Louis d'or* and *écu* were major trade coins of the time, accepted in regional trade, although France briefly shared the reserve currency status with Britain. Despite repeated attempts, France never established a deep, trusted financial system to rival London's, preventing a sustained reserve status during its heyday because it repeatedly suspended its currency's convertibility during wars, losing the trust of holders. Later, hyperinflation during the French Revolution, and then the Napoleonic wars, destroyed French credibility once again, so the Bank of France was only able to stabilize the franc in 1803.

Geopolitically, France used its navy to protect colonies in the Caribbean and India to sustain its wealth, and to project power in European coalitions against Britain. By stationing a fleet in the Caribbean (West Indies), France forced Britain to station 30-40 ships there year-round, which tied down British forces at a huge cost to London. Blockades, like those attempted against Britain during the Napoleonic Wars, aimed at starving rivals, though they were often countered by British superiority because France's blockades were usually ambitious but overstretched. The fiscal return of French naval efforts was trade protection and colonial revenue, but defeats like at Trafalgar (1805) exposed compared to Britain's more disciplined navy.

In short, the French philosophy of supremacy relied on selective engagements and economic attrition. Ultimately, French naval strategy, for all its ingenuity, could not overcome Britain's deeper financial resources and truly global reach. Nevertheless, France perfected asymmetric maritime economic warfare long before the term existed and came closer than any power to bankrupting Britain at sea. Even so, it could never convert its tactical and economic success into lasting sea control because it never solved the problem of maintaining a credible battlefleet during general European war.

British Naval Strategy (1700s–1945)

British naval strategy during the Sterling reserve-currency era was a self-reinforcing cycle of predatory wealth-capture that evolved into a sophisticated global wealth-preservation system. Of all nations, Britain developed the most perfect symbiosis between sea power and wealth generation until the British Treasury, not the fleet, ran out of money.

After the Napoleonic Wars, by the 1890s approximately 60% of world trade was invoiced in the Pound' sterling and the gold sovereign was considered the safest liquid asset and the most widely held coin across the world. Also, the existence of the Royal Navy guaranteed that a bill drawn on London would always be paid because the creditor could always ship the goods.

With this A+ financial standing being the case, while the British pound can be considered the world's reserve currency from about 1815-1914 ("Pax Britannica"), we should extend the historical period we review to analyze how Britain developed such a system. Britain's navy has since passed the position of dominance in 1945–1956 to the U.S. Navy not because the Royal Navy lost a battle, but because the City of London and the British taxpayer could no longer afford to pay for the insurance policy (its navy) that kept the British pound as the world's money.

The British Royal Navy's dominance from the 18th century through World War II essentially employed sea power as the bedrock of economic supremacy and empire. In effect, the Royal Navy was responsible for ensuring Britain's colonial wealth and global influence, so it was essentially the guarantor of Britain's economic lifeline. By 1700, British ships carried 40% of Europe's seaborne trade, and from 1713-1783 the British locked in the first truly global trading empire by seizing the West Indies sugar islands and establishing the North American colonies.

The British philosophy of naval warfare centered on absolute sea control, achieved through superior shipbuilding, rigorous training, and a network of global bases (such as Singapore, Aden and Hong Kong coaling stations) which enabled rapid response to threats. The navy was a central tool of empire that starved rivals of resources via trade disruption when the British deemed it appropriate.

Therefore in the Napoleonic Wars, British blockades strangled France's economy. The Battle of Trafalgar (1805), where enemy lines were broken to annihilate their fleets, cemented British supremacy over France so that Frech colonial trade was annihilated by 1807. The British merchant fleet doubled 1793-1815 while every rival collapsed, leaving Britain with 50% of world tonnage.

Geopolitically, the British Navy enforced colonial rule and trade monopolies, using gunboat diplomacy to coerce smaller nations as seen in the Opium Wars (1839–1860) that forced Chian to pay an indemnity that funded the navy for a decade. During the "Pax Britannica" (1815–1914), the Navy secured trade routes, suppressed piracy, and controlled chokepoints like Gibraltar and the Suez Canal for its own benefit. A consequential peace dividend produced stability in global commerce, which helped foster Britain's industrial dominance in the world. World War I's Battle of Jutland (1916) showcased the "fleet-in-being" concept, where the Grand Fleet preserved British naval power without any decisive victory, which didn't matter because Britain kept the trade routes open.

In WWII, the Navy protected Atlantic convoys, shuttling troops and supplies to Europe and North Africa, ensuring Allied survival against U-boat wolfpacks. Britain's navy was both a weapon and a shield that enforced global trade and colonial rule through a combination of presence, deterrence, and military force when necessary. The Royal Navy's philosophy of supremacy rested on maintaining presence through technological advances like radar and anti-submarine warfare, and global reach with bases enabling power projection from the Caribbean to the Indian Ocean.

Nonetheless, by 1945, economic exhaustion and the rise of U.S. naval power signaled the end of British supremacy, with its Navy transitioning to a supportive role in NATO. Essentially, Britain's empire waned as its naval dominance faded, which offers possible lessons for our modern times. In history it had created a naval-economic model that lasted for two centuries wherein the British navy protected trade, the trade paid taxes and insurance while creating shipping profits, taxes and London loans funded an ever larger navy that in turn protected ever-larger trade. The system finally went broke because the British Treasury went bankrupt after two world wars, and the United States was willing and able to take over the protection role for free under the Bretton Woods / NATO system, at which point Britain passed the baton.

U.S. Naval Strategy (1898–Present)

While the U.S. dollar began overtaking the Pound sterling as the global reserve currency during WWI, once again we must lengthen our examined time horizon to a little before this date. The U.S. Navy's ascent began with the Spanish-American War (1898), and its structure embraced Mahan's vision of a battle fleet designed to secure sea lanes and project influence. From Mahanian expansion it developed into the global economic enforcer, and has reached a point of limits because of the increasing level of U.S. debt service and the ending of the Petrodollar.

Around 1907–1909, the Great White Fleet signaled America's global ambitions and demonstrated the U.S. Navy's capacity for worldwide operations. At this time, the U.S. Navy's philosophy gradually evolved from coastal defense to global trade dominance by prioritizing a carrier-centric fleet (enabling portable air power), amphibious capabilities, and technological superiority.

During WWII, the Navy's island-hopping campaign, led by admirals like Raymond Spruance, defeated Japan through carrier task forces instead of battleships while massive troop transports facilitated amphibious landings from Normandy to Okinawa. The Pacific was basically won by carriers and Marines, which was the carrier revolution.

The Cold War next transformed the U.S. Navy into a global force where carriers served as "floating bases" for power projection from Korea to the Persian Gulf. The Cold War naval strategy now shifted to deterrence rather than forward aggression, with nuclear submarines and carrier battle groups countering Soviet expansion in order to contain Communism. However, the Vietnam-era ops showed the limits of U.S. power projection.

A 1970s agreement between the United States and Saudi Arabia, where Saudi Arabia agreed to price its oil in U.S. dollars in exchange for military support and security guarantees from the U.S., birthed the Petrodollar. This arrangement created a system where OPEC nations priced oil in USD. Their oil sales generated surplus dollars that were then reinvested in U.S. Treasuries, a process that became known as "Petrodollar recycling." The global oil trade created an endless demand for dollars, and resulted in lower U.S. borrowing costs. Through the Petrodollar agreement, the USD secured its role as the main instrument of global currency reserves.

One of the next major developments was that President Reagan's 600-ship Navy and Maritime Strategy (1986) advocated for operations in Soviet waters, namely an offensive forward naval presence in response to the changing geopolitical situation. The Navy also enforced the Petrodollar via Gulf patrols, and the 1991 Gulf War tested its ability at launching precision strikes. While costly, the Gulf War demonstrated U.S. power projection capability and reinforced the security architecture underpinning the global oil trade.

Post-1991, amid multipolar changes the U.S. Navy became the enforcer of "Pax Americana," combating Somali piracy and patrolling critical regions like the South China Sea and Persian Gulf to secure oil tanker routes that underpinned the Petrodollar system. The petrodollar system turned oil – the world's most vital commodity – into a relentless engine of demand for U.S. currency and debt, cementing American financial hegemony. During this time, America's role as global maritime policeman created unprecedented responsibilities, but its naval presence created a "peace dividend" of global trade stability.

The classic gunboat diplomacy of the *Pax Britannica* once used to coerce smaller nations has waned, replaced by a more complex regime of sanctions, carrier group demonstrations, and alliance politics. Modern threats to U.S. naval supremacy include Chinese A2/AD capabilities, Russian submarines, Iranian missile swarms, and smaller drone attacks that make carriers and surface ships increasingly vulnerable. This has resulted in a reliance on standoff strikes and unmanned systems, and is causing a rethink of naval structure.

Going Forward

This is a very abbreviated history of naval warfare through various eras, focusing on a small set of selected principles, while also offering a rare summary of naval evolution according to the necessities faced by the nation offering the world's reserve currency.

Today the dominant naval power is still the United States, but it is being challenged by Russia and China whose cases we will consider in the final chapter. With this short excursion through history finished, we are ready to review the naval commanders highest rated by over 100 experts as the best naval commanders of all time.

CHAPTER 2
A CONSENSUS ON THE TOP 15 NAVAL COMMANDERS

In the previous chapter we traced the evolution of naval warfare and strategy over the last 2,000 years. During that process we mentioned some notable admirals like Nelson or Spruance, but as of yet have not introduced the rankings of top naval commanders.

What I specifically wanted to focus attention upon was how navies and naval strategy was altered by the fiscal reality of the richest nation at the time who depended upon the navy for its revenues. The summary was not meant to focus on the evolution of naval warfare, but to illustrate how navies were used to maintain its economic and geopolitical supremacy in light of technological and fiscal constraints.

The review ended with a short recap of British and then American uses of naval power. Previously, the Pound sterling was the world reserve currency but now the dollar fills that role as the major medium of exchange, although this is rapidly changing.

In *The General's Playbook*, I used sabermetrics to determine a top tier of the most successful military generals of all time (e.g., Napoleon, Caesar). *The General's Playbook* focused on land tactics (ex. flanking, encirclement, divide and conquer, attrition, etc.), while this book focuses on the most successful battle tactics and strategies in sea warfare.

The quandary for this book was objectively coming up with a list of the top tier, most successful naval commanders of all time, ranked by battle success, so that we might study their strategies and contributions to naval warfare. I sought a list of the top fighting naval commanders based on their win-loss records, ability to overcome numerical or technological disadvantages, and their lasting impact on naval strategy.

I turned to AI to help create this list because of its polling abilities. I asked *five* research-based AI engines to search 100 naval historians, naval strategists, and famous naval commanders to compile their own lists of the greatest fighting naval commanders of all time, and then each produced a list of the top 20 ranked by "greatness." By referencing 100 experts for their lists of the top commanders, I found a remarkable degree of consensus on those who were high in battle-winning abilities. By comparing these separate lists I was able to construct a final consensus ranking.

The final ranking synthesizes the consensus of naval historians, strategists, military experts and naval commanders, which as expected identified the same individuals. The farther down one drops in rankings the more opinions diverge as to who is better than another, but as to the top tier of best naval commanders there is a very consistent consensus.

This list, for all its faults and the arguments against it, *definitely collects a top tier of commanders who showed the ability to win major naval battles and decisively influence the course of naval warfare.*

The rankings, however, cannot really tell which admiral is greater than another. It was only used to pull out the top tier of historically great naval commanders recognized for their ability to win naval battles.

Yi Sun-sin's undefeated battle record against overwhelming odds scores higher than Nelson's record so he is often cited as the greatest admiral and tactician by both Western and Eastern historians. However, Nelson's consistent record of winning key battles and revolutionizing naval tactics, particularly at Trafalgar, and his broader geopolitical impact cause many to cite him as the number one admiral of all time. However, Ushakov won far more battles than either. Does it matter who is the best admiral? We just want a list (group) of the top naval commanders so that we might study the methods and commonalities of this group to instill within our own minds.

So, how do you choose between two naval commanders to say who was better? You just want a list of the top tier naval commanders, out of potentially thousands, in order to derive lessons on tactics and strategy from the best. We want to derive our lessons from the cream of the crop rather than argue about who is better than another.

The following "top 10" list is a compiled consensus from naval warfare experts and reflects the most successful naval commanders in history based on their record of winning battles and shaping the course of maritime warfare. The criteria for inclusion is achieving decisive victories in major naval battles, the ability to win against superior forces or in adverse conditions, and an enduring impact on naval history and doctrine because of their influence on naval warfare:

1. Yin Sun-sin, 23-0 win-loss record
2. Horatio Nelson, 8-1 win-loss record
3. Tōgō Heihachirō, 3-0 win-loss record
4. Themistocles, 2-0 win-loss record
5. Michiel de Ruyter, 6-1 win-loss record
6. Raymond Spruance, 3-0 win-loss record
7. Chester W. Nimitz, 5-0 win-loss record
8. Andrew Cunningham, 3-0 win-loss record
9. David Farragut, 3-0 win-loss record
10. Hayreddin Barbarossa, 5-1 win-loss record

I do not like including Themistocles in this list as one of the greatest naval commanders of all times because he offers limited lessons for modern commanders. Themistocles was not the overall naval commander of the Allied Greek fleet at the Battles of Artemisium and Salamis in 480 BC. However, he was indeed the commander of the Athenian contingent (the largest and most important part of the fleet) and effectively the driving strategic force behind both battles. His remarkable strategy won one of the most significant naval battles of western civilization that altered its destiny, and this consistently earns him inclusion by experts voting their opinions for a list of greatest naval commanders.

If I personally had to rank these naval commanders in terms of greatness – their ability to win naval battles (e.g., overall record and consistency), win against the odds (e.g., numerical or technological disadvantages), outstanding tactics that influenced generations (e.g., innovations in maneuvers, amphibious operations, or psychological warfare), and the unique, instructive lessons they – I would reorder this list as follows:

1. Yi Sun-sin
2. Horatio Nelson
3. Hayreddin Barbarossa
4. Michiel de Ruyter
5. Chester W. Nimitz
6. David Farragut
7. Raymond Spruance
8. Tōgō Heihachirō
9. Andrew Cunningham
10. Themistocles

A few words must be said about the exclusion from this list of the Chinese Admiral Zheng He, the famous Chinese explorer and fleet commander who commanded history's largest wooden fleet in seven epic voyages. He was extremely successful in expanding Chinese influence but excluded for lack of pitched battle experience. The list was derived to highlight commanders who might offer tactical naval lessons for combat-focused warfare.

Nevertheless, Zheng He's career offers important lessons in soft power diplomacy and intimidation through size. He offers a great example of pre-modern naval power projection through overwhelming scale. His fleet – a floating city – embodied a "sword and silk approach" of military greetings or diplomacy because it carried either cannons or a promise of prosperity to the nations he visited.

Rounding out this list we have a further five admirals, whose inclusion once again prioritizes their ability to win battles, especially against the odds, using tactics or innovations that influenced subsequent generations, and the unique lessons they offer. I disagree with these rankings provided by our experts, but this is their conclusion:

11. Francis Drake , 4-1 win-loss record
12. Andrea Doria, 4-1 win-loss record
13. Thomas Cochrane, 5-0 win-loss record
14. Don Álvaro de Bazán, 40-0 win-loss record
15. Fyodor Ushakov, 43-0 win-loss record

Several famous naval commanders should be mentioned who were consistently cited by our experts as being the in the "great" category, yet did not make it into this list of the *top 15* greatest of all time. We can only speculate as to why.

Isoroku Yamamoto was a brilliant Japanese WWII strategic planner whose Pearl Harbor attack demonstrated bold carrier aviation tactics, but his Midway assault was a catastrophic failure due to overcomplex plans with poor contingency planning. This probably dropped him from being in the list of the top 15. His carrier-centric strategy definitely influenced WWII naval doctrine, but his tactics were not as influential or odds-defying as other naval commanders.

William Halsey was an aggressive and dynamic U.S. naval leader in the Pacific during WWII. However, credit for many of his largest victories must be shared with other admirals (notably Spruance). Several of his operational decisions, including his response to Japanese decoy tactics at Leyte Gulf, have also been criticized as costly and his impulsiveness (e.g., leaving Samar unprotected) nearly cost the battle. He probably did not make the upper list because of these issues. Furthermore, his carrier operations were innovative but risky and while bold, his tactics were not as uniquely influential at the doctrinal level as those of Yamamoto or Ushakov.

Maarten Tromp was a superb Dutch admiral of the 17th century who pioneered line-of-battle. He won against odds in Anglo-Dutch wars and won important victories. However, his period's tactics, while brave and often innovative for their time, offer few unique lessons and do not resonate as directly with modern naval doctrine. Additionally, his influence did not shape global naval thinking in the same way as most of these other naval figures.

John Jellicoe is best known for his command that preserved the British fleet at the Battle of Jutland (WWI), but he failed to deliver a truly decisive result despite having overwhelming superiority. Historians consider his approach conservative and overly cautious, and his tactics largely defensive while lacking innovation or odds-beating flair. This limits their inspirational value for teaching aggressive or creative naval warfare. He was also criticized for delaying the introduction of the convoy system against German U-boats, which could have reduced Allied losses earlier.

Karl Dönitz was a master of submarine (U-boat) warfare. He created the "wolf pack" tactics that won early Atlantic victories, but were an overall loss. Furthermore, his tactics were strategically significant but largely attritional, focusing on economic strangulation rather than fleet actions. His command decisions rarely involved direct fleet battle or decisive tactical maneuvers that are broadly applicable as general lessons across naval warfare contexts.

Afonso de Albuquerque was another admiral often cited as belonging in the top tier. He was a pioneering Portuguese figure in conquering Indian Ocean ports against the odds, and used bold amphibious tactics. However, his victories often came from applying overwhelming force or local superiority, and the distinct tactical concepts he introduced are less transferable to modern naval instruction compared with the global strategic innovation of Yamamoto or the fleet action lessons of Ushakov.

As we proceed downwards in rankings, the number of suggested officers explodes, and it becomes very difficult to even group naval commanders into clearly defined tiers of better and worse. Therefore, I ended this list with the top 15 naval commanders of all time, paralleling the rankings I compiled for land generals, and you will find my own determination of the top five best naval commanders after their biographies, which follow.

CHAPTER 3: YIN SUN-SIN (1545–1598)

Admiral Yi Sun-sin (1545–1598) of Korea's Joseon Dynasty was an unparalleled naval commander who fought 33 battles and engagements, winning all 33 (23 of them major fleet actions). He fought in the Imjin War (1592–1598) against Japan where the Japanese aimed to conquer Korea, viewing it as a vital stepping stone for a planned – though arguably aspirational – campaign against Ming China. Yi sunk hundreds of enemy ships with inferior technology (Korea had no cannon-armed sailing ships) but never lost a ship under his command. He almost always fought against numerically superior fleets and regularly won not because of luck but due to his superior naval tactics. His tactical genius, innovative ship designs, and disciplined leadership transformed the Korean navy into an indomitable force, and his battles achieved the strategic salvation of Korea when every rational calculation said it was impossible.

Admiral Yi Sun-sin's remarkable success was built upon a bedrock of meticulously developed strategic principles and an unwavering commitment to preparation. Yi's most effective strategies and tactics – the innovative Turtle Ship design, the Crane Wing formation, tidal warfare through superior knowledge of coastal terrain and currents, precision firepower, and unbreakable morale – are some of the reasons why he secured his legendary victories. His methods provide a masterclass in how to defeat a stronger opponent through superior wit, fire control, and force of will as he regularly crushed numerically superior fleets with minimal losses.

The Turtle Ship (*Geobukseon*) – A Revolution in Naval Warfare

Yi's signature innovation was the Turtle Ship (*Geobukseon*), often cited as the world's first armored warship – 250 years before the *Monitor* or *Glorie*. Its design is detailed in Yi's war diary, and a separate description was later produced by his nephew. It was a fully enclosed, iron-plated battleship built to dominate close engagements. Its design consisted of an iron-plated roof armed with spikes that deterred enemy boarding, which was a major vulnerability in the traditional sea combat of the time and a formidable advantage against the Japanese who preferred close-quarters melee combat where samurai swarm enemy decks.

Armed with multiple cannons positioned on all sides, the Turtle Ship functioned as a powerful mobile artillery platform and a heavily armored ramming vessel. Its impenetrable armor and overwhelming firepower allowed it to sustain heavy damage while inflicting devastating blows. By combining his turtle ships with cannon-armed heavy boarding ships, Yi created a combined-arms fleet that could destroy Japanese ships at range and then annihilate survivors in close combat.

The Turtle Ship's dragon-headed prow, possibly used to emit smoke from internal braziers, served primarily as a terrifying psychological weapon to disorient Japanese crews. With only three to five built due to high costs, each was a force multiplier, spearheading assaults and breaking enemy lines.

At the Battle of Sacheon (1592), Yi debuted the Turtle Ship, sinking 13 Japanese ships without losing a single Korean vessel. It rammed enemy lines, fired broadsides, and withdrew under cover, a tactic repeated at Dangpo (1592), where 21 Japanese ships were destroyed. In the climactic Battle of Noryang (1598), the Korean fleet, likely led by its remaining Turtle Ships, helped destroy over 200 Japanese vessels.

Essentially, the Turtle Ship's design disrupted prevailing naval doctrines, which is a common characteristic of the strategies used by our leading naval commanders. This design, making boarding virtually impossible, fundamentally changed the nature of naval engagement in favor of ranged artillery. Its maneuverability, powered by 80 oarsmen and two sails, allowed rapid repositioning in tight coastal waters. By targeting enemy flagships, Turtle Ships sowed chaos, collapsing command structures.

The Crane Wing (*Hakikjin*) Formation – A Tactical Masterstroke

Yi's signature *Hakikjin* (Crane Wing) formation was a U-shaped ambush that lured enemies into a kill zone of deadly crossfire. It was designed to envelop the enemy fleet, allowing Yi's ships to deliver concentrated broadsides from multiple angles. The fleet formed a wide crescent, lured the Japanese in, and then closed the wings while ships attacked the center. By feigning retreat, Yi would entice Japanese fleets into narrow straits where his ships hit them with crossfire from three sides, sealing escape routes. Hence, the formation combined the perception of vulnerability with the strategy of encirclement, and thus destruction.

At the Battle of Hansando (1592), Yi's 56 ships faced a force of 73 Japanese vessels and the Crane Wing formation played a decisive role in annihilating the Japanese fleet. Yi deployed the Crane Wing to sink or capture 59 enemy ships while losing none of his own. The decisive victory crippled Japanese naval power in the south, severing Japanese supply lines and undermining their entire land campaign. In the Battle of Myeongnyang (1597), with only 13 ships against 133 Japanese vessels, Yi used the strait's currents and the Crane Wing to funnel enemies into a slaughter, sinking 31 vessels in a great underdog victory.

The Crane Wing was a perfect trap for the Japanese, exploiting their aggressive pursuit and inability to coordinate under sudden, multi-directional attack. The formation's flexibility allowed adjustments based on geographical terrain. Yi's disciplined gunners also held their fire until enemies were fully committed, maximizing the impact of his cannons. The U-shape of the formation ensured that no Japanese ship could flank or retreat without facing devastating volleys.

Mastery of Tidal and Coastal Warfare

Yi never fought in open seas unless it favored him, preferring shallow coastal waters, narrow straits, and tidal zones. He made use of shallows where Japanese deep-keel ships were vulnerable (so that he could ground Japanese deep-draft ships), tidal straights with unpredictable currents that hampered enemy fleet cohesion and maneuvers, and islands and shallows in order to conceal and ambush. His coastal moves naturally demanded intimate knowledge of the Korean coastline, its intricate network of islands, narrow straits, and, crucially, its treacherous currents and tides.

Whereas a general uses terrain on land, Admiral Yi's coastal tidal knowledge allowed him to choreograph maritime terrain for his battles. At the Battle of Okpo (1592), Yi trapped 26 Japanese ships against an outgoing tide, sinking them with concentrated fire as they floundered in the tidal backwaters. Japanese ships, which were designed for open-water piracy, struggled in Korea's complex coastal geography that was unknown to them and became prey to his maneuvers.

Yi's knowledge of tidal patterns allowed precise timing, as at Myeongnyang, where strong currents scattered Japanese formations. Myeongnyang's success hinged on tidal shifts that Yi predicted. Yi's *panokseon* ships, flat-bottomed and stable, maneuvered easily in the shallows, outpacing enemy vessels. In the Battle of Busan (1592), he also lured 100 enemy vessels into coastal shallows, destroying 70 by exploiting their immobility.

Decisive Firepower Discipline – The "One Shot, One Kill" Doctrine

Admiral Yi was an advocate of outstanding artillery accuracy under pressure. His fleets employed a "one shot, one kill" philosophy, prioritizing artillery precision over volume. Korean cannons, like the *Jija* and *Hyeonja Chongtong*, outranged Japanese arquebuses, and were able to devastate enemy ships before boarding became feasible. The Japanese naval tactics favored grappling and boarding enemy ships for melee combat, but Admiral Yi's strategy emphasized inflicting maximum damage at range to cripple or sink enemy vessels before they could close the distance for boarding.

Yi also trained gunners to target enemy flagships first in order to decapitate the enemy command and cause panic and disarray. At Dangpo (1592), Yi's 26 ships sank 21 Japanese vessels by targeting their flagship, killing commander Kurushima Michiyuki. At the Battle of Noryang (1598), which was the final major naval engagement of the Imjin War, Yi's forces again focused on attacking leadership vessels to collapse the enemy's structure and morale. His precision firepower decimated 200 enemy ships where burning Japanese hulks became obstacles that amplified the chaos and damage. Yi was fatally wounded at this battle, but it was a resounding victory, crippling the Japanese retreat and securing their ultimate withdrawal from Korea.

Turtle Ships shielded reloading *panokseon*, allowing sustained volleys without exposure. Unlike European broadside tactics, Yi's gunners aimed for critical hits, and Japanese accounts of the battle noted the devastating and precise effect of Korean cannon fire that we might liken to a missile barrage.

Unbreakable Morale, Leadership and Care for His Men

Yi's greatest asset was his men's loyalty, forged because he led from the front, shared hardships, and inspired courage. As Alfred Thayer Mahan observed, a navy's strength flows from the will of its commander. Before Myeongnyang, with a mere 13 ships against 133 (and 200 support vessels), Yi rallied his fleet saying, "If you seek death, you will live; if you seek to live, you will die." (*Nanjung Ilgi*, 1597). This shows his character and resolve, for Yi had just returned to command after unjust imprisonment and torture almost to the point of death, being spared the death penalty but demoted to the rank of a common infantry soldier. During his absence the Korean Navy's 150 warships with 30,000 had been decimated so that he only had 13 ships at his disposal.

With only 13 ships, he chose the Myeongnyang Strait – a narrow channel notorious for its treacherous, unpredictable currents and whirlpools – to negate the Japanese numerical advantage. He meticulously studied the tides, timing his attack to coincide with a shift in the current that would first aid his smaller fleet and then turn against the larger, less maneuverable Japanese vessels. This terrain created a natural bottleneck, forcing the Japanese ships into a confined space where their numerical advantage became a hindrance.

Myeongnyang's improbable triumph, with two Korean casualties against 18,000 Japanese, also rested on his personal courage, unwavering leadership, and his men's morale. Korean success at the pivotal Battle of Noryang at the end of the Imjin War, despite Yi's death, also reflected his enduring influence. His final words at Noryang, "Do not announce my death; the battle is at its height," are a testament to his spirit and dedication to his nation.

Yi is known to have paid his sailors and marines out of his own pocket when the court withheld funds. He refused to abandon wounded men but carried them off under fire. Japanese accounts describe Korean sailors fighting to the death because "they would rather die than disappoint Yi Sun-sin." Yi's presence on the flagship, often under fire, galvanized his crews. His strict but fair discipline – punishing cowardice but rewarding bravery – built trust. After his demotion and reinstatement following the Battle of Chilchonryang (1597), where his successor lost most of the fleet, Yi rebuilt the navy from 12 ships to 53 in months, a tremendous feat. As stated, his death at Noryang (1598), which was kept secret to maintain morale, helped ensured its victory. There the Korean navy destroyed the Japanese fleet, sinking about 300 of their 500 ships, to prevent future invasions.

Yi Sun-sin's Enduring Legacy

Admiral Yi Sun-sin's unparalleled record was forged by a seamless fusion of technological ingenuity, tactical artistry, disciplined gunnery, and unbreakable morale. The Turtle Ship revolutionized naval design, the Crane Wing maneuver mastered positioning, tidal warfare weaponized geography by turning the coastlines and narrow seas into weapons, precision fire maximized lethality, and his leadership style fused strategic clarity with shared purpose.

Tōgō Heihachirō, perhaps Japan's greatest admiral, confessed, "It may be proper to compare me with Nelson, … but not with Yi Sun-sin. He is too great to be compared to anyone." Admiral Yi did not merely win battles but created a system for victory. He reshaped naval doctrine by proving that a smaller, determined, and well-led force could defy the odds and annihilate a superior enemy, so his tactics offer us a blueprint for winning against superior foes by outthinking one's enemies. In an era of drones and missiles, Yi's principles remain relevant. Admiral George Ballard of the Royal Navy said, "Yi is the most perfect example of a sea generalship in history." Alfred Thayer Mahan stated, "Yi proved that genius, not numbers, commands the sea."

CHAPTER 4:
HORATIO NELSON (1758–1805)

Horatio Nelson (1758–1805) remains an enduring figure in naval history, admired and celebrated for his audacious tactics, inspirational leadership, and decisive victories against numerically superior foes. Unlike his contemporaries, who clung to the rigid conventions of 18th century line-of-battle warfare where fleets would maneuver to approach enemy ships in parallel lines, Nelson revolutionized naval engagements with innovative strategies that went against this formalism. Instead, he prioritized speed, surprise, and an uncompromising pursuit of annihilation rather than the target of simply damaging the enemy.

His victories at the Battles of Cape St. Vincent (1797), the Nile (1798), Copenhagen (1801), and Trafalgar (1805) didn't just win battles; they shattered French naval power for a generation and imposed a new, aggressive template for naval warfare. What separated Nelson from his contemporaries was a ruthless mindset: he saw the rigid line not as a rule, but as a target.

To understand Nelson's genius, we must first grasp what he overthrew. Eighteenth-century naval tactics were dominated by the Fighting Instructions, rigid rules for fleet tactics that prescribed formal line-of-battle engagements. Fleets typically approached each other in parallel lines, exchanging broadsides at long range until one of the sides withdrew. These battles were typically indecisive affairs where one of the opponents achieved tactical superiority without strategic resolution. The strategic objective for navies was often less about achieving a decisive victory than it was about avoiding losing since this would entail the immense cost and difficulty of replacing capital ships and their experienced crews.

Nelson's genius was practical, not theoretical. He viewed the vaunted *Fighting Instructions* as an overly rigid recipe for engagements and deliberately cast them aside. While admirals like Villeneuve or Calder adhered to centralized control, it was Nelson's flexibility from rejecting the rules and becoming unpredictable that allowed him to exploit enemy weaknesses and win decisive battles. His relentless offense was encapsulated in his maxim, "Never mind maneuvers, always go at them." Alfred Thayer Mahan observed, Nelson's "genius lay in adapting principles to circumstances" rather than following prescribed formulas.

Nelson understood that the Royal Navy's advantages – superior seamanship, faster gunnery, and disciplined crews – were wasted in formal engagements that played to enemy strengths. British sailors spent the vast majority of their time at sea and drilled endlessly in seamanship. He understood that British crews, who could fire a broadside every 90 seconds – nearly twice as fast as their opponents – were wasted in long-range duels. Nelson wanted to take advantage of such British advantages, so his tactical revolution transformed the predictability of linear warfare into the deadly improvisation of close combat, and he rested on creating chaos where British training could dominate.

Breaking the Line: Disrupting Enemy Cohesion

Nelson's signature tactic, breaking the line, involved a perpendicular assault driving his ships through the enemy's linear formation to split it into smaller, manageable segments. This was a revolutionary tactic that defied the standard line-of-battle doctrine of parallel engagements to maximize broadside firepower.

By cutting directly into the opposing fleet, Nelson aimed to divide it, induce confusion, deny those ships mutual support and coherent command, and allow his own ships to rake their opponents fore and aft – an angle of attack that inflicted the most severe damage and could not be returned by broadsides. British ships could deliver devastating broadsides down the length of enemy vessels, and it is estimated that raking volleys caused triple the damage of standard broadsides.

Nelson notably employed this signature tactic in three decisive engagements: the Battles of Cape St. Vincent, Battle of the Nile, and Trafalgar.

As a commodore under Admiral Jervis, at the Battle of Cape St. Vincent Nelson broke out of the line of battle without direct orders to engage three Spanish ships with his lone vessel, ultimately capturing two of them. This wasn't recklessness, but a calculated aggression to intercept the Spanish fleet that turned a potential stalemate into a British victory. The British captured four ships of the line, which was an outstanding result considering the size disadvantage. Admiral Jervis not only excused Nelson but praised him, and shortly thereafter he was made a Knight of the Bath, and then promoted from Commodore to Rear-Admiral of the Blue.

At the Battle of the Nile, Nelson repeated this disruption on a small portion of the French line. With the French fleet anchored in Aboukir Bay and their landward side presumed safe, Nelson split his force – sending half between the anchored French line and the shore. The resulting crossfire devastated the French and trapped Napoleon's army in Egypt. The French lost 11 of 13 ships of the line.

At Trafalgar, facing a Franco-Spanish fleet of 33 ships to his 27, Nelson divided his fleet into two columns, sailing them perpendicularly into the enemy's extended line splitting it into three segments, isolating the enemy's center and rear and leading to a decisive victory by destroying the enemy's cohesion.

The reason that cutting the line worked so well is that it exploited the fundamental weakness of linear tactics, which was their dependence on maintaining formation integrity. Cutting the line severed enemy communications because signal flags became obscured in the smoke and chaos of battle, so the strategy disrupted enemy command. It basically neutralized an enemy's ability to coordinate its fleet.

Cutting the line also enabled raking fire. British ships could fire broadsides down the length of enemy vessels, targeting vulnerable bows and sterns. In fact, N.A.M. Rodger estimates that raking volleys caused up to three times the damage of standard broadsides (*The Command of the Ocean*, 2004). Brian Lavery notes that British ships often fired three broadsides for every two by French or Spanish opponents (*Nelson's Navy*, 1989), whose crews were largely inexperienced and plagued by poor gunnery. British crews, trained for rapid gunnery, excelled in the ensuing close-quarters melee from cutting the line.

The Nelson Touch: Bold, Unconventional Maneuvers

The "Nelson Touch," a term Nelson himself coined, was a philosophy that defined his leadership style: empower your captains, inspire your men, and attack with ferocity. Nelson had a preference for rapid, unexpected (unconventional) attacks that defied traditional caution. One of the lessons we can derive from his battles is that rapid, unpredictable maneuvers can neutralize superior forces.

Two examples of Nelson's unconventional maneuvers were at Copenhagen and Trafalgar. At Copenhagen, Nelson ignored a signal to withdraw when facing a fortified Danish fleet, and continued pressing the attack that ultimately forced a Danish surrender when a formal approach would have failed. This is the battle where – when ordered to disengage – he famously placed his telescope to his blind eye and said, "I really do not see the signal!"

At Trafalgar, the perpendicular approach of Nelson's fleet in two columns shocked the Franco-Spanish fleet, leading to a chaotic melee that prevented the enemy from forming a cohesive defense so that British seamanship prevailed. His sudden unexpected maneuvers commonly caught enemies – who were typically expecting formal maneuvers – off guard. Speed and surprise were Nelson's tactical weapons to disrupt enemy plans.

To mitigate the drawback of attacking head-on, which exposed the leading British ships to devastating raking broadside fire, Nelson had ordered his ships at Trafalgar to make "all available sail," including stunsails. This daring maneuver allowed them to close the gap with the Allied fleet far more quickly than was customary. Nelson had assessed in a calculated gamble that the French and Spanish gunners would struggle to fire accurately at his fleet from a moving platform, especially with the heavy swell.

The victory at Trafalgar effectively crushed French and Spanish naval power, thwarting Napoleon's invasion plans and securing British sea supremacy for the subsequent hundred years. This dominance allowed Britain to maintain its strategy of blockading France, effectively starving it of resources and disrupting its trade routes.

The Melee Strategy: Close-Quarters Annihilation

Nelson focused on annihilation over attrition – aiming to destroy enemy fleets entirely rather than just damage them (which was the aim of contemporaries) – and this underscored a need for overwhelming strikes that ensured lasting strategic gains.

His tactic to achieve annihilation was to force a melee – a confused, close-quarters fight – where British seamanship and gunnery could dominate. The chaos of a melee disrupted enemy command chains, making it difficult for admirals to issue orders, and typically led to decisive outcomes because they prevented the enemy from retreating, which was in line with Nelson's relentless pursuit of achieving decisive annihilation. Nelson wanted the battle to degenerate into a massive, uncontrollable, ship-to-ship pell-mell brawl at pistol-shot range (50–200 yards) as quickly as possible because he believed that in such chaos British crews, British gunnery, and British fighting spirit would always destroy any opponent.

Therefore, Nelson deliberately sought to transform naval battles into melee close-range brawls, where British discipline and gunnery could dominate in the close-quarters combat. This favoritism for melees contrasted with a French and Spanish preference for long-range engagements that would preserve their fleets.

An example of Admiral Nelson's preference for melee was seen at Trafalgar, where after breaking the line, Nelson's fleet engaged in an aggressive devastating melee at point-blank range, capturing or destroying 22 of 33 enemy ships without losing a single British vessel. Nelson himself was unfortunately struck by a French musket ball and died three hours later, but not before learning his victory was complete.

Psychological Warfare and Leadership from the Front

Nelson's leadership was as critical as his tactics. Paul Kennedy notes that "Nelson's men fought harder because they believed in him" (*The Rise and Fall of British Naval Mastery*, 1976), so their high morale became a force multiplier in battles. Nelson put a big emphasis on crew welfare ensuring ample and healthful provisions, promoting fair treatment, and consistently rewarding merit. He inspired fierce loyalty by sharing plans openly with his officers, and lead from the front on *Victory* where his signal and visible presence galvanized his fleet. He shared risk and hardship with his crews; he also cultivated a deep camaraderie among his officers. Sir Julian Corbett, a prominent British naval historian, acknowledged Nelson's unique leadership style noting that Francis Drake's gift for personal leadership was "unmatched until Nelson's time."

Although he fought in an era defined by rigid adherence to the line of battle, Nelson empowered his captains to act independently once battle commenced, trusting their independent initiative. Trusting subordinates allowed them to freely adapt in chaos, which gave his fleet adaptability. Nelson created a decentralized command structure, revolutionary for an era of "follow the rules," where initiative thrived.

Nelson cultivated what he called his "*Band of Brothers*" – a cadre of captains who understood his thinking and carried it out without hesitation. He stimulated thought on tactical questions, encouraging frank expression of ideas from them in a continuous educational process of discussion that would lead to them thinking alike about tactical principles, which helped produce a unity of action across his fleet. This degree of trust and understanding enabled the decentralized execution that complex operations demand. By briefing his captains extensively, ensuring they understood both tactics and intent, Nelson paved the way for his plans to be executed flawlessly, and the prior briefings ensured captains like Collingwood acted decisively.

This decentralized command empowered captains to act without waiting for signals because they understood intent and could adapt quickly to changing circumstances, and exploit fleeting opportunities, without detailed orders. This delegated autonomy stood in stark contrast to the more centralized, and often less effectively executed, command systems of his French and Spanish opponents.

By briefing his "Band of Brothers" on his intent rather than issuing a rigid written plan as was customary, Nelson created a primitive but highly effective form of mission command, allowing his 15 captains to adapt decisively so that his fleet transformed into a pack of independent killers. Richard Harding praises his model of empowered subordinates as "a blueprint for modern naval leadership" and Mahan, in *The Life of Nelson* (1897), praises Nelson's trust in his subordinates as "a model for modern fleets."

Why Nelson's Tactics Surpassed His Peers

Nelson's numerous victories resulted from several factors, one of which was the fighting readiness of the two fleets (British and French) since Nelson's fleet spent months at sea maintaining blockades, while French ships remained in port. The Royal Navy's emphasis on seamanship training, competitive gunnery practice, and aggressive tradition created crews capable of executing complex maneuvers under fire. Nelson leveraged this foundation through tactics that maximized British advantages while exploiting enemy weaknesses. One might say that Nelson amplified the existing strengths of his fleet and transformed them into decisive advantages.

While fighting with *essentially the same technology as his opponents*, Nelson maximized the potential of existing systems. British ships were better maintained, crews better trained, and commanders more experienced in actual combat operations. Nelson's example teaches that technology alone does not determine victory.

Nelson's tactical innovations transcended their technological context because they addressed timeless principles of leadership, decision-making, and organizational behavior. His genius lay not in kinetic actions rather than in following rules for he possessed an intuitive grasp of the moment to abandon convention and strike. Admiral Nelson's strategy of annihilation rather than attrition ensured lasting strategic gains. His approach became institutionalized throughout the Royal Navy, creating a culture of aggressive professionalism. Michael Duffy notes, "Nelson didn't just win battles; he reshaped naval warfare."

By combining unconventional aggressive tactics (breaking the enemy line and concentrating force) with inspirational leadership, systematic planning with tactical flexibility, and personal courage with strategic vision, Nelson became a transformative figure who redefined what became achievable in naval combat. He created a template for naval excellence that remains relevant today despite centuries of technological change.

The lesson of Nelson's battles is not to replicate his specific tactics but to emulate his approach: understand your strengths and enemy weaknesses well enough to take calculated risks in daring the unexpected.

CHAPTER 5:
TŌGŌ HEIHACHIRŌ (1848–1934)

Admiral Tōgō Heihachirō (1848–1934) was the architect of Japan's emergence as a maritime power during the Russo-Japanese War (1904–05). His masterful execution of established tactics – especially "crossing the T" – along with a revolutionary focus on long-range gunnery and radio intelligence, allowed Japan to defeat a larger Russian fleet, cementing Japan's status as a global naval power. His triumph at the Battle of Tsushima (May 27–28, 1905), where he annihilated Russia's Baltic Fleet, is often cited as one of history's most lopsided naval victories, definitively established Japan's naval ascendancy.

Tōgō's strategies were shaped by his training at Britain's Royal Naval College, Greenwich (1871–78), where he studied advanced naval tactics and technology. Upon returning to Japan, Tōgō became a central figure in modernizing the Imperial Japanese Navy (IJN), integrating British doctrines with Japanese samurai culture.

Tōgō's forged the Imperial Japanese Navy into a force capable of outmatching larger foes by combining cutting-edge Western technology and the doctrine of the Royal Navy with the relentless discipline of the samurai tradition. His modernization efforts ensured that the IJN was equipped with modern battleships like *Mikasa* and trained crews. His experience in the Sino-Japanese War (1894–95) further honed his skills, preparing him for the Russo-Japanese War. However, it was his exposure to Western navies that really gave him the greatest strategic edge over Russian commanders.

The "Crossing the T" – Perfected Naval Enfilade

The hallmark of Tōgō's naval genius was his use of "crossing the T," a maneuver where his fleet moved perpendicular to the advancing Russian line. This positioning enabled his ships to fire full broadsides, concentrating his firepower, while the Russians could only bring limited forward guns to bear. In this formation, Tōgō's disciplined crews maintained a tight formation, delivering deadly accurate salvos while weathering only limited return fire.

At the Battle of Tsushima (1905), Tōgō executed this tactic twice, collapsing the Russian formation completely by sinking 21 Russian ships and capturing 7 with over 4,000 sailors killed. This was the only time in history that an entire battleship fleet was destroyed in a single engagement. The Japanese losses were minimal – three destroyers lost and fewer than 120 killed.

At the Battle of the Yellow Sea (1904), Tōgō attempted to cross the T, but the Russians managed a fighting retreat. The battle ended inconclusively when a lucky Japanese salvo killed Vitgeft and caused confusion on the Russian flagship.

Long-Range Gunnery Dominance

Admiral Tōgō emphasized fire control systems and long-range gunnery training, drilling Japanese crews relentlessly. Consequently, at Tsushima Japanese gunners achieved 3–4 times the Russian hit rate, igniting enemy magazines and sinking battleships in minutes. His accuracy was also partly due to using advanced British rangefinders that enabled accurate salvos at distances the Russians could not match.

Additionally, Tōgō used *shimose* shells, filled with high-explosive picric acid, which caused fires and explosions upon impact. These high-explosive shells gave the IJN yet another gunnery edge, especially with its rapid, disciplined salvo firing that maintained a constant barrage. The Admiral's flagship *Mikasa* alone fired over 100 accurate shells into the Russian line in the opening minutes of the Battle of Tsushima.

This fusion of superior technology and fanatical training gave Tōgō's fleet the ability to cripple enemies at ranges where they could not effectively respond. At Tsushima, gunnery sank multiple Russian battleships. During the Port Arthur blockade, Tōgō's bombardments eroded the Russian defenses.

Night Torpedo Attacks

It was also Tōgō's practice to unleash destroyers and torpedo boats under cover of darkness to weaken the enemy before daytime battles, and to finish off crippled Russian vessels. These relentless night attacks prevented any Russian recovery and ensured no damaged enemy ship would escape to fight another day. At the Battle of the Yellow Sea, for instance, the night attacks prevented Russian ship repairs. These night attacks were a systematic "mopping up" operation, a final, ruthless phase of his battle plan.

Night attacks at Tsushima sank at least six additional ships. In fact, half of the Russian losses at Tsushima came from night torpedo attacks. They caused early damage to several Russian vessels, drove the Russians into predictable courses, deprived them of rest, and created the chaos that he exploited the next morning when he crossed the T.

Decisive Fleet Concentration

Unlike his Russian counterparts who split their fleets (the Pacific Squadron and the Baltic Fleet), Tōgō maintained a unified Combined Fleet, ensuring that he could bring overwhelming local force to bear at a decisive point. This principle echoed Mahan's principle of "concentration of force," which Tōgō felt was more important than maneuvering individual squadrons independently.

Tōgō's practice of never dividing his fleet ensured local superiority against Russia's divided forces in every engagement. Unified forces allowed Tōgō to avoid being defeated piecemeal and to dictate the terms of engagement so that he could exploit Russian exhaustion and disarray.

Tōgō's genius included strategic patience – the discipline to avoid battle until conditions guaranteed his advantage. At the Battle of Tsushima, the entire Combined Fleet engaged the Russians. At the Battle of the Yellow Sea, concentration prevented Russian breakouts.

Intelligence and Reconnaissance

Tōgō prioritized intelligence, using scouts and wireless telegraphy, which was pioneering for its time, to track enemy movements. Accurate intelligence enabled Tōgō to choose the battlespace so that he could exploit Russian exhaustion after their fleet's 18,000-mile journey. He also employed wireless telegraphy for real-time command and control.

Before the Battle of Tsushima, Japanese scouts – using advanced radio and visual reconnaissance – accurately tracked the Russian Baltic Fleet across the East China Sea, allowing Tōgō to position his fleet for the decisive T-cross maneuver that would intercept the Russian force at precisely the moment it entered the Tsushima Strait.

His use of radio ensured real-time coordination, and Tsushima marked the first modern naval battle where radio played such a critical role.

Psychological Warfare and Command Disruption

Tōgō understood that warfare was as much psychological as kinetic. Before Tsushima, he hoisted the morale-boosting Z-flag aboard *Mikasa*, famously declaring: "The fate of the Empire depends on this battle. Let every man do his utmost." This invocation of national destiny before the fight electrified his sailors.

During the battle, he targeted Russian flagships early to cripple the enemy's command-and-control. When the *Knyaz Suvorov* – Admiral Rozhestvensky's flagship – was sunk after Tōgō's salvos, the rest of the Russian fleet lost cohesion. The sinking decapitated the Russian command, shattering morale and causing the fleet's collapse. At the Battle of Port Arthur, the sustained pressure of the Japanese navy broke Russian morale.

Tōgō's annihilation of the Russian Baltic Fleet at Tsushima (38 vs. 38 ships, with Russia losing 34) remains one of history's most complete naval victories. Admiral Tōgō's tactics became models for a new era of naval warfare, and he is sometimes referred to as the "Admiral Nelson of the East" for delivering a Tsushima-like decisive victory.

He was more than a brilliant tactician; he was the architect of Japan's naval doctrine, the symbol of its ascent, and the strategist who secured its place among the great powers. Admiral Tōgō Heihachirō is famous for saying. "You may win without fighting, but you will not win without strategy." His destruction of the Russian fleet at Tsushima shocked the world, signaling the arrival of Japan as a major power.

CHAPTER 6:
THEMISTOCLES (c. 524–459 BC)

Themistocles (c. 524–459 BC), an Athenian statesman and the strategic mastermind behind the Greek naval victory at Salamis that defended Greece against the Persian Empire. His innovative strategy had three pillars: first, building a formidable trireme fleet; second, luring enemies into confined waters for an ambush; and having shop crews master the *diekplous* maneuver. He employed psychological deception against the enemy enabling the Battle of Salamis (480 BC), one of the most decisive single sea victories in Western history. His victory at Salamis was not just tactical but existential for the independent Greek city-states, fundamentally altering the course of European history and the fate of Western civilization.

Themistocles' naval strategies overcame a great numerical inferiority to secure Greece's survival against Persia and establish Athens as a maritime power. While other Greek commanders relied on hoplite warfare, Themistocles foresaw the power of sea dominance. His most fundamental strategic act was convincing Athens to build a formidable *trireme* fleet, financed by a windfall of silver from the Laurion mines, which gave Greece its essential naval tool. Athens built 200 *triremes* (the wooden wall prophesized by the Oracle of Delphi), creating a fast, agile fleet manned by trained citizen rowers (170 oarsmen, 20 sailors and 14 marines), outmatching Persian conscripts.

His crowning achievement – the Battle of Salamis (480 BC) that used these *triremes* – stands as one of the most brilliant naval victories of all time, where Themistocles defeated a Persian fleet three times larger through superior tactics, terrain mastery, and psychological warfare. This was the largest naval battle of the ancient world.

The *Trireme* Revolution – Building Athens' Naval Power

As mentioned, in 483 BC Themistocles persuaded the Athenian assembly to use profits from the Laurion silver mines to construct 200 *triremes*, which were fast and agile warships powered by three tiers of oars. Themistocles' foresight in building this fleet countered Persia's ability to supply its vast army by sea, which was a critical vulnerability.

The *trireme's* design – light, fast, and equipped with a bronze ram – enabled devastating ramming tactics that outmatched the heavier, less agile Persian galleys. However, because the *trireme* was a precision instrument, it required coordination, timing, and skill that necessitated training. It was a weapon of maneuver, not mass, and was ideally suited to Greece's narrow straits and coastal waters. It was the weapon that allowed a polis like Athens to punch far above its weight, transforming naval combat from a boarding melee into a contest of maneuver and skill.

The trireme was only as good as its crew. Unlike Persian conscripted rowers, Athens' citizen rowers (*thetes*) were highly motivated trained extensively, and shared in the political fate of their city. Themistocles' emphasis on citizen rowers, which was a type of political participation, fostered the buildup of discipline and morale. Altogether these superior conditions translated directly into the agility, coordination, and relentless aggression that won at Salamis.

The Salamis Trap – Luring the Enemy into a Kill Zone

When the Persian king Xerxes I invaded Greece with the mightiest armada the world had yet seen, many Greeks argued for retreat or open-sea battle. Themistocles instead advocated for an ambush in the narrow straits between Salamis Island and the mainland. He understood that Xerxes was eager for a decisive victory, prone to overconfidence, and his towering pride made him vulnerable to a well-crafted lie.

Themistocles' masterstroke at Salamis was luring the Persian fleet into the narrow straits between Salamis and the mainland, which negated their numerical advantage. By sending his servant Sicinnus to Xerxes with a false message of Greek disarray and planned retreat, this subterfuge prompted the Persians to attack in confined waters that were essentially a slaughter pen. The deception worked perfectly. Flush with overconfidence, Xerxes ordered his vast fleet into the straits at dawn, sailing directly into Themistocles' carefully laid trap.

Of the ~370 Greek ships at the Battle of Salamis, Athens contributed nearly 180. The narrow straits prevented the massive Persian fleet – perhaps 600 to 800 ships – from deploying effectively in the cramped waters where their numerical advantage became a liability. The confined space caused chaos among the Persian fleet because their larger vessels collided together and lost formation.

Greek *triremes*, more agile and manned by trained crews, maneuvered tightly in the waters and executed precise ramming attacks, sinking 200–300 Persian ships while losing only about 40. Although the rams used on both sides could cripple ships, stricken vessels continued to float rather than sink so after the initial clash the battle broke down into melees where the Greek hoplite's equipment, training and motivation began to show in the hand-to-hand combat. Xerxes' own brother was one of the casualties of battle. The Battle of Salamis (480 BC) saw the Greek fleet, under nominal Spartan command but effectively led by Themistocles, decisively defeat the Persians, halting their invasion of Greece.

Xerxes retreated to Asia with much of the fleet, leaving a large army in Greece which was defeated the following year at Plataea. The Persian naval threat, however, remained significantly diminished.

The *Diekplous* Maneuver – The *Trireme's* Deadliest Tactic

Themistocles' fleet had perfected the *diekplous*, a maneuver where *triremes* broke through gaps in enemy lines in single file, turned, and rammed vulnerable enemy ship sides or rears. The maneuver – breaking through enemy lines to turn and ram vulnerable flanks or shear off oars – demanded precision that only Athens' trained rowers could deliver.

The *trireme's* speed and agility allowed quick penetration and withdrawal, thereby avoiding Persian boarding tactics. On the other hand, the typical strategy of the heavier, more rigid Persian ships (which fared poorly in close quarters) was to row towards an enemy, board them and then kill the rowers. As a result of its design and movements, the *diekplous* was able to disrupt enemy formations at Salamis by creating a chaos that shattered Persian cohesion, thus exposing weak points for attack.

Naval Fortifications and Strategic Withdrawals

Themistocles' strategic patience – knowing when to avoid battle – preserved Athens' naval strength. He fortified Piraeus (located southwest of Athens), creating a secure naval base, and ordered retreats to conserve forces for decisive engagements. Fortifying Piraeus ensured a safe harbor for the Greek fleet, enabling sustained operations. Post-Thermopylae (480 BC), Themistocles orchestrated Athens' evacuation and fleet concentration at Salamis. Post-Salamis, the fortifications at Piraeus solidified Athens' maritime empire.

Themistocles' Enduring Legacy

Themistocles' genius was multifaceted: the visionary who built the fleet, the psychologist who deceived a king, the tactician who forced mastery of the *diekplous* maneuver, and the strategist who successfully manipulated the Persians into committing their fleet at the exact moment in the place he wanted. Although he was unappreciated in his final years and died in exile, his naval strategies transformed Athens into a maritime power, saved Greece and shaped Western civilization.

CHAPTER 7:
MICHIEL DE RUYTER (1607–1676)

Michiel de Ruyter (1607–1676) was the Netherland's greatest admiral and is still seen as a symbol of the Dutch Golden Age. A man of humble origins from Vlissingen, de Ruyter rose through merit to become the supreme admiral of the Dutch fleet and won over 30 major engagements against formidable English, French, and Spanish fleets during the turbulent Dutch Golden Age. His rise from humble Zeeland sailor to the republic's supreme admiral cemented his status as a folk hero – a man of the people who triumphed through sheer grit and genius. He is considered by some historians to be the greatest naval commander of the 17th century.

De Ruyter's innovative tactics, disciplined command, and adaptability secured the Dutch Republic's maritime dominance during the Anglo-Dutch Wars. His innovative tactics – disciplined line-of-battle, aggressive breaking of the enemy line, feigned retreats, squadron-based command, and combined arms – enabled the Dutch Republic to maintain maritime dominance despite being outnumbered by their foes. De Ruyter was considered cautious and prudent, yet his strategies consistently outmaneuvered enemies and remain studied for their modern relevance.

Disciplined Line of Battle (*Linietaktiek*)

Though not the first to employ the "line of battle," de Ruyter was its most consistent and effective practitioner in the 17th century. The *linietaktiek* – aligning ships bow-to-stern in a tight line – concentrated broadside fire and allowed the admiral to maintain control of his fleet. De Ruyter intensely trained his crew in the maneuver, which brought a new level of discipline and order to the Dutch fleet as at the time naval battles were characterized by a chaotic nature. He saw the line as a way to enable his fleet to act as a single, coordinated entity in battle.

De Ruyter's relentless drilling in the line-of-battle forged a new standard of disciplined, concentrated firepower that became the standard in European naval warfare. This approach required strict discipline to prevent ships from breaking ranks to pursue personal glory, which would expose the entire line to destruction – a weakness the English were eager to exploit. Therefore, de Ruyter prevented isolated ship engagements by emphasizing strict formation. In many battles, de Ruyter directed his attack toward the enemy's middle squadron, often engaging their flagship directly.

By aligning all guns for simultaneous broadsides, the line-of-battle formation maximized firepower and simply overwhelmed enemy ships. De Ruyter's emphasis on discipline ensured that fleet cohesion was maintained under fire, which allowed the Dutch navy to act like a synchronized machine. He concentrated broadside fire at key decisive points, and flexible squadron commands allowed responses adaptive to the situation. His line required strict discipline, but within that framework, he trusted his squadron commanders to use their initiative in a way that supported the overall battle plan.

The English often relied on aggressive melees, but de Ruyter's methodical approach minimized Dutch losses while maximizing damage to the enemy navy. By keeping ships in a tight line to maximize firepower and minimize losses, he was able to consistently triumph even against larger fleets.

As examples, de Ruyter's disciplined line-of-battle formation broke the Swedish blockade of Copenhagen at the Battle of the Sound (1658), sinking or capturing 12 ships while relieving Danish forces. At the Four Days' Battle (1666), his masterful command and line-holding techniques during four days of continuous fighting and multiple changes of wind turned initial setbacks into a grueling strategic victory, forcing the English fleet into retreat. At the Battle of Schooneveld (1673) and Texel (1673) he similarly prevailed against larger fleets, and in these four battles he inflicted 3-10 times the losses he received almost entirely because his lone-of-battle was maintained with a discipline no contemporary admiral could equal.

Crossing the T: Precision Firepower

Decades before Horatio Nelson, de Ruyter perfected cutting across an enemy's line ("crossing the T"), allowing his fleet to rake enemy ships with devastating full broadsides into the enemy's vulnerable brow or stern areas while receiving minimal return fire. This tactic disrupted enemy fleet formations by splitting them into isolated segments, while also targeting command ships because that often decided battles.

By crossing the enemy's bow or stern, de Ruyter exploited the limited firing arcs of 17th-century ships. His gunners' precision, honed through rigorous training, ensured devastating volleys at this vulnerable moment of passing through the line.

De Ruyter used this maneuver at Battle of Lowestoft (1665) where, even though this was a Dutch defeat, his rearguard actions using this tactic saved much of the fleet. At the Battle of Solebay (1672), his fleet successfully crossed the Anglo-French line, splitting their 90-ship fleet and sinking three vessels, thwarting an invasion.

A signature element of de Ruyter's tactics was his use of the rearguard, which he often commanded himself even though the position was the most dangerous. In this position, he could absorb the brunt of the enemy's attack, protect the main body of his fleet, and then tactically launch a devastating counter-attack at the perfect moment. He perfected (and used repeatedly) a specific rearguard tactic that allowed a smaller or damaged Dutch fleet to disengage from a superior enemy while continuing to inflict heavy punishment and avoiding the usual disaster of a running fight.

De Ruyter commanded the rearguard during the Four Days' Battle, one of the longest and most grueling naval battles in history where de Ruyter's disciplined lines held against 56 English ships with 84 Dutch vessels delivering coordinated broadsides over four days, sinking 10 English ships. On the second and third days he used the tides and shoals to his advantage, rallied his scattered fleet and turned the tide of the battle. In terms of losses the battle was a tactical draw, but the Dutch forced the English to retreat, which a stunning outcome given the circumstances.

Deception and Feigned Retreats

De Ruyter's frequently used feigned retreats to lure enemies into traps, exploiting their overconfidence and unfamiliarity with Dutch waters. His signature weapon was a deliberate, large-scale feigned retreat that was a rehearsed performance designed to pull a stronger enemy fleet out of formation. No other 17th century admiral ever used large-scale feigned retreats so repeatedly and successfully.

Using his deep knowledge of Dutch coastal terrain, by pretending disarray he drew opponents into its shoals, currents or narrow channels, limiting enemy mobility, and then counterattacked. Special colored flags or lanterns told every captain exactly when the retreat was to become a counter-attack, and at that time his entire fleet would fall on the enemy with local superiority. Military historian Russell Weigley (*The Age of Battles*) notes that de Ruyter's use of feigned retreats was more sophisticated than most land generals of his era.

Two examples where he used such deception were at the Raid on the Medway (1667) where he tricked the English into thinking he had withdrawn, whereas he then sailed up the Thames and Medway, burning or capturing much of the English fleet (13 ships) including the English flagship *HMS Royal Charles*. At the Battle of Schooneveld (1673), de Ruyter twice lured a superior 90-ship Anglo-French fleet into shallow waters, neutralizing their numbers and sinking six ships with no Dutch losses.

Squadron-Based Command: Decentralized Initiative

De Ruyter permanently divided his fleet into three semi-independent squadrons (van, center, rear), each led by a trusted flag officer (Tromp, Evertsen, Van Nes, etc.). Unlike rigid hierarchical fleets, de Ruyter empowered squadron commanders were given full tactical freedom with an overall intent, so they were empowered to act independently on their own initiative rather than wait for centralized commands. As a result, his subordinate commanders could adapt mid-battle without waiting for orders. His standing orders were:

- "Act as if you were the admiral yourself."
- "When you see a chance to damage the enemy then take it without waiting for my signal."

- "Support the nearest sister squadron in danger, even if it means leaving your station."

This proved vital in chaotic battle engagements, especially against larger enemy fleets. Unlike rigid English and French fleets, his decentralized command enabled squadron leaders to exploit sudden opportunities without waiting for orders like targeting damaged enemy flagships or any gaps in enemy lines. Such flexibility allowed the Dutch fleet to rapidly adapt to changing conditions, while rigorous training ensured an alignment with de Ruyter's intent. Historian Jaap R. Bruijn (*The Dutch Navy of the Seventeenth and Eighteenth Centuries*) argues, "This trust in subordinates was key to Dutch success" against larger foes.

At Texel (1673), for instance, squadron commanders' maneuvers to adapt in real time prevented encirclement by a larger Anglo-French fleet, protecting the Dutch Republic from invasion.

Combined Arms: Naval Gunfire & Boarding Actions

While many admirals of the period favored either long-range artillery duels (English) or tried immediate close-quarter boarding actions, de Ruyter trained his sailors to do both and transition between the two as the battle demanded, thus merging the two strategies. He blended long-range cannon fire with selective boarding so that he might first weaken enemies with gunnery as a preparation before capturing ships intact, which was a versatility that unnerved foes used to the rigidity of fixed doctrine. His men, drilled relentlessly, were therefore effective whether behind cannons or fighting on enemy decks.

This balanced approach maximized the damage caused to enemies as well as his resource gains from winning ships, setting him apart from one-dimensional naval rivals. The initial gunnery of low, smashing broadsides had the objective of killing or wounding 30-50% of the enemy crew and destroying its artillery before any attempts to board. This reduced the risks for his boarding parties, which secured captured ships in minutes.

As examples, at the Battle of Plymouth (1652) he used precise cannon fire to disable English ships and afterwards sent boarding crews to finish the job, capturing three English vessels. At the Battle of Dungeness (1652), his method enabled him to capture multiple ships intact for the Dutch fleet, which we should consider valuable prizes.

De Ruyter inflicted 3-5 times the casualties he took in most battles. He also captured more capital ships in a single war than any admiral before Nelson, and did it with a smaller fleet.

Why de Ruyter's Tactics Succeeded

De Ruyter died from a cannonball striking his leg at the Battle of Agosta in 1676, his last breath taken in service to his country. It is recorded that de Ruyter's French opponents (he fought against the French in the Franco-Dutch War) honored him by firing salutes as the ship carrying his body to the Netherlands passed their fleet.

As a devout Christian, de Ruyter led prayers before battle, and faith had sustained him through the four days of combat during the Four Days' Battle in 1666. His piety was well-known among his men, a source of morale, and his men nicknamed him *"Bestevaêr"* (grandfather). Historian John Keegan remarks, "De Ruyter's faith was his secret weapon" (*The Price of Admiralty*).

De Ruyter's genius lay in his tactical innovations (line of battle, crossing the T), deceptive maneuvers (feigned retreats, ambushes), decentralized command (trust in subordinates) and adaptive combat style (merging firepower and boarding). These tactics, such as his perfection of the line, allowed him to defeat larger fleets, while his versatility ensured adaptability across diverse battlefields and his trust in his captains produced a flexibility to strike where it hurt most. De Ruyter was also involved in the founding of the Netherland's Marine Corps.

His many winning tactics and track record explains why he is revered as the savior of the Dutch Republic in its golden age and a foundational figure in the history of naval warfare.

CHAPTER 8:
RAYMOND SPRUANCE (1886–1969)

Admiral Raymond Ames Spruance (1886–1969), often called the "quiet warrior," was a master of calculated strategy. As a key figure under Admiral Chester W. Nimitz, Spruance orchestrated decisive victories in the Pacific Theater of World War II that dismantled Japan's naval dominance and secured Allied control of the Pacific. He was the victorious fleet commander at the two most decisive carrier battles of the war: Midway (1942) and the Philippine Sea (1944), which secured the invasions of Midway and Saipan (1944) respectively. Spruance's Midway victory, executed under Nimitz, showcased tactical brilliance, sinking 4 Japanese carriers with fewer ships.

Spruance's reputation among naval historians and strategists is that of the "best tactical mind of any naval commander in the Pacific theater." His signature tactics – precision ambushes, disciplined carrier operations, a defensive-offensive balance, night surface actions, situational awareness, leadership under pressure, and technological integration – set him apart from predecessors like Horatio Nelson, who relied on bold, aggressive fleet engagements.

Known for his calm judgment, cool-headed precision and calculated restraint, Spruance contrasted with the fiery aggression of peers like William "Bull" Halsey, yet his prudence produced the triumphs at Midway (1942) and the Philippine Sea (1944), delivering devastating blows to Japan's navy with minimal American losses.

The Offensive Principle of Calculated Risk

One of Spruance's signature tactics was the principle of *calculated risk*. This doctrine, articulated by Admiral Nimitz and executed by Spruance, called for inflicting maximum damage on the enemy while minimizing any exposure to counterattack. In other words, achieve maximum effect with minimum risk.

At Midway, Spruance's decision to launch a full-scale air attack at extreme range at the earliest possible moment based on a *calculated guess* of the Japanese location after scout planes found some of the enemy carriers, epitomized this approach. He recognized that surprise and concentration of force were paramount – if he could strike first, he could destroy them, but if discovered, their superior force would prevail.

Spruance's willingness to launch early, risking aircraft fuel shortages, was a bold but calculated deviation from standard doctrine. This decision, made in consultation with his staff, resulted in the destruction of three Japanese carriers in a single blow and turned the tide of the Pacific War. As naval historian Craig L. Symonds observed, "The actual battle of Midway was a mixture of Japanese hubris, American luck, and outstanding leadership by Spruance."

Situational Awareness

Spruance's acute situational awareness was a critical tactic that allowed him to adapt tactics mid-battle. At Midway, he adjusted strike timings based on real-time scout reports; at the Philippine Sea, he repositioned his fleet based on radar and air reports to counter Ozawa's Plan; at Iwo Jima, he adapted air strikes to shifting *kamikaze* threats.

Unlike peers who relied on rigid plans, Spruance synthesized intelligence reports (from Station Hypo codebreaking), weather, fleet status and other conditions into decisive fleet placements, using them to make better real-time battle decisions.

The night of June 19-20, 1944 at the Philippine Sea is universally regarded as the finest example of pure battlespace awareness in carrier warfare history.

Precision Ambushes by Leveraging Intelligence

Another signature tactic was the precision ambush. Leveraging intelligence, Spruance carefully positioned his fleet for devastating ambush strikes against critical Japanese targets, recognizing that his fleet would usually be going against numerically superior forces. The supreme example of the precision ambush was Midway, the turning point of the Pacific War.

Decrypts revealed Admiral Yamamoto's plan to invade Midway, allowing Spruance to position Task Force 16 (25 ships including *Enterprise* and *Hornet)* against Japan's 145-ship group. Holding carriers northeast of Midway until Japan's air groups were committed ensured surprise. Launching full air groups to attack, despite the fuel risks, ended up overwhelming Japanese defenses. On June 4, dive bombers from the *Enterprise* and *Yorktown* sank three Japanese carriers (*Akagi, Kaga, Sōryū*) in five-minutes; the fourth, *Hiryū*, was sunk later that day by a follow-up strike. Spruance lost only the *Yorktown* in the battle.

Spruance's willingness to launch at extreme range for this ambush, risking his aircraft running out of fuel, was a bold deviation from standard naval doctrine. This decision for the daring raid, made in consultation with his staff, took into account the principle of calculated risk, and resulted in the destruction of three Japanese carriers in a single blow with the fourth later in the day. His decisions and the consequential battle turned the tide of the Pacific War.

Spruance's command style focused on mission clarity. At Midway, once the Japanese carrier threat was neutralized, Spruance resisted the temptation to pursue the retreating enemy into a night battle, where he "calculated" that his forces would be at a disadvantage. He withdrew after victory to avoid night battle with Yamamoto's battleships, protecting Midway and preserving his remaining strength, which was a decision vindicated by postwar analysis.

At the Battle of the Philippine Sea (June 1944), often called the "Marianas Turkey Shoot," Spruance commanded the Fifth Fleet to protect the Saipan invasion against Admiral Ozawa's Mobile Fleet. Informed by decrypted signals, he positioned Task Force 58 (TF 58), which included 15 carriers and supporting ships, west of the Marianas in a defensive arc, with battleships screening the landings.

Refusing aggressive westward pursuit to meet the Japanese head-on, Spruance let the Japanese attack first, deploying hundreds of Hellcat fighters in a massive air screen to protect his fleet. On June 19, as Japanese strikes launched, U.S. radar-enabled intercepts destroyed ~600 enemy planes and three Japanese carriers (*Hiyo, Shokaku, Taiho*) with only ~130 American losses, crippling Japan's carrier-based air power. His defensive positioning had turned the engagement into a devastating ambush.

That night, prioritizing the invasion over chase, Spruance moved east, thwarting Ozawa's shuttle-bombing plan from Guam. A late-afternoon counterstrike on June 20 sank more ships, but distance limited total destruction. This victory secured the Marianas while permanently crippling the Japanese naval air arm, and effectively ended the Japanese Navy's ability to conduct large-scale carrier operations for the remainder of the war. This ensured the ultimate success of the Central Pacific campaign. This calculated restraint not only secured the Marianas, but enabled B-29 bombings of Japan, and led to Prime Minister Tojo's fall. It proved that Spruance's defensive-offensive balance conserved forces while luring enemies into traps.

A Defensive-Offensive Balance

Spruance balanced defensive positioning with opportunistic offensives. This set him apart from aggressive peers like Admiral Halsey, who often pursued risky chases. Spruance prioritized strategic objectives (e.g., protecting Midway) over seeking fleet battles, which went against Mahan's doctrine of seeking decisive engagements. He conserved his fleet for decisive moments, avoiding unnecessary losses while waiting to capitalize on enemy errors.

By maintaining a defensive posture until intelligence confirmed an advantage, Spruance acted to preserve his fleet's strength rather than pursue risky chases, while attempting to lure enemies into traps. His calculated restraint ensured sustainable campaigns, and helped U.S. forces outlast Japan's overstretched navy.

His success at the Battle of the Philippine Sea was an example of prioritizing the mission (protecting the Saipan invasion force) over the opportunity for a high-risk engagement to chase and destroy the Japanese fleet, despite criticism from some aviators for not pursuing a more aggressive attack. By keeping Task Force 58 near Saipan, Spruance was able to ensure the Saipan invasion's success, avoid Yamamoto's trap, and save thousands of lives. By waiting for Japan to exhaust its air arm before counterattacking, Spruance also ensured a lopsided victory. His Task Force 58 launched 1,200 sorties over two days, destroying Japan's air arm with coordinated strikes while still maintaining a defensive screen.

Spruance's comprehensive thinking ensured that the *primary objective* – the success of the amphibious landing – was never compromised for the sake of tactical glory, and by forcing the Japanese to attack his prepared defenses he engineered an air ambush that permanently crippled the Japanese naval air arm, securing the Marianas and ensuring the ultimate success of the Central Pacific campaign.

Night Surface Action

Spruance's night actions deployed cruisers and destroyers for surprise attacks. Unlike Japanese admirals who preferred daylight engagements because they relied on visual spotting, Spruance leveraged the use of radar under darkness to disrupt enemy plans. Spruance integrated radar, aircraft, and anti-aircraft systems into a unified fighting system, which was revolutionary in solidifying the naval shift to air-centric warfare.

Spruance's night-fighting training emphasized torpedo salvos to disrupt enemy formations and radar-directed gunfire from battleships. The results of such training showed at Midway, where Spruance's destroyers conducted night torpedo runs to harass Japanese stragglers, a tactic that minimized the chances of a counterattack. At Guadalcanal (November 1942), Spruance supported night surface actions (under subordinates), sinking Japanese transports by using new SG radar to guide attacks.

Though not at Leyte Gulf, his doctrines influenced the decisive night action at Surigao Strait (October 1944) where U.S. forces annihilated Japan's southern group.

Logistical Strangulation of Fortified Island Bases

Spruance's island campaigns focused on isolating Japanese garrisons via submarines and air raids, and so he bypassed fortified bases but still weakened them without costly invasions. Carrier raids neutralized enemy bases to enable amphibious landings. Spruance also used submarine patrols to sink Japan's merchant fleet and help starve its war machine through economic strangulation.

Key operations that demonstrated this were at Truk Raid (February 1944), where carrier strikes rendered Japan's "Gibraltar of the Pacific" impotent without invasion. At Iwo Jima and Okinawa (1945), Spruance's naval bombardments softened Japanese defenses, minimizing invasion ground losses.

The Spruance Temperament: Disciplined Leadership

It has often been said that Spruance's rejection of risk-laden pursuits preserved the American fleet. For instance, at Midway, Spruance withdrew after sinking Japan's carriers, avoiding a night battle with Yamamoto's battleships, and at Philippine Sea, he resisted chasing Ozawa's decoy carriers, protecting Saipan.

Spruance's unemotional, analytical leadership style under pressure ensured that his decisions aligned with strategic goals, and it not only inspired confidence in his men but ensured disciplined execution. He demonstrated a leadership style quite different than Admiral Halsey's, but delivered devastating victories at the Battles of Midway, Philippine Sea, and Iwo Jima through precision, restraint, and strategic foresight. His tactics dismantled Japan's navy and paved the way for the Allied victory in the Pacific.

Some historians rank Spruance among the greatest admirals, stating that his understated brilliance outshone flashier peers. He wove intelligence, technology, and a profoundly patient aggression into a new form of warfare: one that conserved forces until decisive moments, ambushed with precision, defended with overwhelming force, leveraged technology for tactical superiority and logistically strangled the enemy's capacity to fight.

His ability to secure lopsided victories with minimal losses remains one of the most remarkable records in U.S. naval history. Many of his strategies set new standards for modern naval warfare, as has his practice of employing cold, analytical decision-making in tactical calculation (no impulsive moves). He is also remembered for his focus on strategic priorities over seeking fleet battles (and glory-seeking doctrines), his strategic patience and unassuming leadership under pressure, and the practice of bringing overwhelming firepower to situations to secure naval victory.

CHAPTER 9:
CHESTER W. NIMITZ (1885-1966)

Few naval commanders have shaped the course of a war as decisively as Admiral Chester W. Nimitz (1885–1966), whose leadership as Commander-in-Chief of the U.S. Pacific Fleet during World War II turned the tide against the Imperial Japanese Navy. Nimitz orchestrated a campaign of relentless pressure that systematically dismantled Japanese naval power, redefining warfare at sea in the process.

Born in Fredericksburg, Texas, Nimitz graduated from the U.S. Naval Academy in 1905, excelling in mathematics and submarine warfare. After the devastating Pearl Harbor attack that shattered the U.S. fleet, he transformed a battered fleet into a dominant force that crushed the Japanese Imperial Navy, securing victories in battles like Midway (1942), the Philippine Sea (1944), and Leyte Gulf (1944).

His signature strategies – carrier-centric warfare, island-hopping, submarine warfare, decentralized command and intelligence exploitation – set him apart from predecessors like Horatio Nelson, who relied on aggressive fleet engagements, and peers like Isoroku Yamamoto, whose overcomplex plans faltered. Historian Samuel Eliot Morison opinioned, "Nimitz's leadership was the linchpin of Pacific victory."

At the heart of Nimitz's warfare approach was the principle of calculated risk which was, as already explained, associated with Admiral Spruance. Facing a Japanese fleet that had dominated the Pacific, Nimitz understood that he could not defend everywhere. Therefore he sought to concentrate superior force at decisive points, even if it meant violating standing orders or exposing other areas to risk.

For example, in the run-up to Midway, Nimitz assembled all available American carriers and positioned them, along with land-based aircraft and submarines, to ambush the Japanese *Kido Butai*. This daring concentration of force, executed at "Point Luck" northeast of Midway, allowed the U.S. to spring a devastating trap and turn the tide of the war. In Midway, the odds were unfavorable but acceptable given the intelligence.

A New Paradigm: Carrier-Centric Warfare

Nimitz was a submariner by training but recognized that warfighting platforms evolve, and that aircraft carriers were the future of naval warfare. This perspective marked a paradigm shift from cherished battleship-centric doctrines, which dominated both American and Japanese thinking before WWII. Nimitz pioneered the coordination of carrier task forces, grouping aircraft carriers with escorts (destroyers, cruisers) to project air power across vast distances.

The carrier-centric doctrine as the fleet's primary offensive weapon allowed Nimitz to dominate the Pacific. Carriers provided unmatched flexibility, projecting the power of long-distance air strikes without exposing the naval fleet to concentrated enemy fire, and without relying on forward bases. This went far beyond the range of battleship guns. He therefore changed the naval focus from rigid battleship formations to carrier task forces combining carriers, cruisers, destroyers, and submarines, and pioneered mobile "air-sea" strike groups that resembled floating air bases.

Carrier task forces allowed Nimitz to strike from far beyond enemy range, maximizing their reach, while protecting his carriers with layered defenses that included fighter screens. This countered Japan's larger naval fleet by enabling him to concentrate air power (on carriers) at critical points.

Consequently, Nimitz constructed his carrier task forces as mutually supporting ecosystems of defense and attack, combining fighters, dive-bombers, and torpedo bombers with radar and anti-aircraft screens. His key naval victories where his strategy triumphed were the Battles of Midway (1942), Philippine Sea (1944), and Leyte Gulf (1944).

At the Battle of Midway, his carriers *Enterprise*, *Hornet* and the hastily repaired *Yorktown* executed a meticulously planned ambush, sinking four

Japanese carriers – *Kaga*, *Akagi*, *Soryu*, and *Hiryu* – in a single day, shattering Japanese naval superiority and turning the tide of the war. In that moment, Japanese naval supremacy was irreversibly shattered. In the Battle of the Philippine Sea, U.S. carriers destroyed over 600 Japanese aircraft, securing U.S. air superiority. At Leyte Gulf, the forces under Nimitz's overall command engaged a divided Japanese fleet across three fronts, sinking four carriers and multiple battleships in the war's largest naval battle. And at Okinawa (1945), Nimitz's carrier groups (Task Force 58) supported amphibious landings with relentless air strikes, neutralizing Japanese *kamikaze* threats.

Nimitz didn't just adopt new technologies (e.g., carriers, radar, and anti-aircraft systems); he fused them into a new kind of combat system that outpaced the Japanese reliance on battleships. This was revolutionary in WWII, and under Nimitz's command naval warfare shifted from gun-based to air-based combat. Nimitz was outnumbered by the Japanese, but his rapid adoption of advanced technology often offset Japan's numerical advantages. Radar and anti-aircraft guns enhanced his defensive capabilities, while carriers enabled long-range strikes that were able to overwhelm Japan's outdated fleet.

A Defensive-Offensive Balance

Unlike the aggressive doctrines of admirals like Nelson or Farragut, Nimitz balanced defensive positioning with opportunistic offensives, which were essentially targeted against Japan's larger navy. By maintaining a defensive posture until intelligence confirmed an advantage, Nimitz preserved his fleet while setting traps for overextended enemies. He waited for decisive opportunistic moments, avoiding reckless pursuits while capitalizing on enemy errors.

Examples: At Midway, Nimitz held his carriers back until Japanese forces committed to Midway, then launched a devastating counterstrike. His restraint avoided Japan's trap. At the Philippine Sea, Nimitz positioned his fleet defensively to absorb Japan's air attacks, then launched overwhelming air strikes that destroyed 430 Japanese aircraft. At Coral Sea, Nimitz's defensive deployment protected Australia using limited forces to check Japan's advance, which set the stage for Midway.

Integration of Air-Sea Logistics & New Technologies

Nimitz perfected the art of sustaining naval offensives over long distances. He mastered underway replenishment, an innovation that kept fleets at sea for months by refueling and resupplying them on the move. Endurance at sea provided U.S. forces with the possibility of greater operational reach and surprise.

The sustained pressure across the Central and Western Pacific from 1942–45 relied on floating logistics – a combination of tankers, ammunition, and repair ships coordinated by Nimitz's staff to sustain the fleet.

Nimitz's habit of adopting technology to the fleet also ensured tactical superiority over the Japanese. He integrated all platforms and emerging technologies – carriers, submarines, radar, and anti-aircraft systems – into his tactics, outpacing Japan's reliance on battleships while shifting naval warfare from gun-based to air-based combat. His radar-equipped carriers, for instance, used anti-aircraft guns that neutralized *kamikazes,* and carrier air strikes, guided by radar, sank Japanese ships at Leyte Gulf and Midway.

Intelligence-Driven Ambushes

One of Nimitz's greatest assets was his use of signals intelligence, particularly codebreaking, to position his smaller American fleets to exploit Japanese weaknesses in planning, coordination and overconfidence through ambushes, thus neutralizing their numerical superiority. Unlike earlier admirals who relied on scout ships or open-sea engagements, Nimitz exploited decrypted Japanese communications (e.g., JN-25 code) to anticipate enemy movements. This was so valuable that John Keegan observed, "Nimitz's codebreaking was the equivalent of an extra fleet" *(The Price of Admiralty).*

Nimitz leveraged ULTRA intelligence (decrypted Japanese messages) to give the U.S. a strategic edge via its window into enemy plans. Signals intelligence allowed his staff to predict Japanese movements and intentions with remarkable accuracy, and he made use of this. Intelligence essentially gave the U.S. a critical edge in its battlefield operations. This approach was groundbreaking in the 1940s, as it shifted naval strategy from being reactive to proactive, enabling precision fleet positioning and preemptive strikes.

For example, the U.S. Navy's Station Hypo, led by Joseph Rochefort, decrypted Japanese JN-25 codes, revealing plans for the Midway attack, which enabled a devastating counter-ambush. Before Midway, Japanese messages referred to a target codenamed "AF." Nimitz authorized a clever ruse – Midway's garrison sent an unencrypted message claiming their water desalination plant had failed. When HYPO intercepted a Japanese report stating "AF is short on water," Nimitz knew the attack's destination.

Nimitz trusted in Rochefort's decoding of Japanese plans for an attack on Midway, and avoided micro-managing the fight from Pearl Harbor by issuing orders to Admirals Spruance and Fletcher to "Inflict maximum damage on enemy by employing strong attrition tactics." The boundaries of the engagement were: "In carrying out the task assigned ... you will be governed by the principle of calculated risk ... the avoidance of exposure of your force to attack by superior enemy forces without good prospect of inflicting greater damage than you receive." The overarching strategic goal was to *preserve* the American carriers, as losing more carriers than the enemy would give Japan undisputed command of the sea in the Pacific. They were told to strike hard but pull back if the risk became too great.

The U.S. positioned carriers (only 25 U.S. ships against Japan's 145) at a location northeast of Midway Island dubbed "Point Luck." This location was out of the range of Japanese search planes but put the U.S. carriers on the flank of Nagamo's approach, allowing the American planes to strike first. Nimitz had planned an ambush, relying on long-range PBY reconnaissance planes flying from Midway to locate the Japanese carriers first, which would then trigger the American air strike from the concealed carrier position. Midway's victory hinged on knowing Japan's plans in advance.

Similarly, at the Battle of the Coral Sea, intelligence alerted Nimitz to Japan's invasion force headed toward Port Moresby (New Guinea), allowing him to ambush them in the Coral Sea with two carriers (*Lexington, Yorktown*), using the Solomen Islands as a cover. The battle halted Japan's advance, sinking one carrier (*Shoho*). At the Battle of the Philippine Sea, decrypted signals also enabled Nimitz to anticipate Japan's carrier movements so Admiral Raymond Spruance positioned his 15 carriers to devastate hundreds of Japanese aircraft in the "Great Marianas Turkey Shoot."

Island-Hopping (Leapfrogging)

Rather than costly frontal assault on every Japanese island stronghold, Nimitz pioneered a campaign of strategic bypass where he captured critical islands to serve as stepping stones to Japan while isolating others by cutting off their supplies via submarine blockades. Instead of underplaying subs, thanks to his submarine experience he knew how to leverage them early to help starve the Japanese war machine. By denying Japanese strongholds the ability to be resupplied, Nimitz basically rendered them strategically irrelevant to the broader military campaign.

Known as "island-hopping," this strategy allowed the Allies to seize specific enemy airfields, project power, and cut Japanese supply lines without exhausting their forces and wasting countless lives on costly assaults. This strategy was critical to conserving American resources, and accelerated the Allied advance across the Pacific while grinding Japan down.

Nimitz's island-hopping strategy bypassed heavily fortified Japanese strongholds like Truk (the main Japanese naval base in the South Pacific theatre, located in the Caroline Islands of the Federated States of Micronesia), but captured key bases (Rabaul, Tarawa, Iwo Jima) to extend Allied reach while isolating enemy garrisons. The strategy was also seen in the Philippines Campaign where U.S. Forces retook Leyte while avoiding Mindanao, and in the Gilbert and Marshall Islands campaigns the U.S. forces seized Tarawa, Makin, Kwajalein, and Eniwetok, establishing airfields and anchorages.

Submarine Warfare & Economic Strangulation

Under Nimitz's command, the U.S. submarine force waged a silent but catastrophic war on Japanese merchant shipping, especially targeting oil tankers to cripple the IJN's mobility. By war's end, Nimitz's submarine campaign had sunk more than 1,300 Japanese ships, including over 55% of Japan's merchant fleet and eight aircraft carriers, which devastated Japan's economy.

Having few major natural resources of its own, Japan relied heavily on imported raw materials from Southeast Asia, which was the whole reason for their expansion into the region. The economic strangulation of submarine warfare basically crippled Japan's war effort, as seen in continuous Pacific patrols from 1943 to 1945, and was as critical as fleet battles because this continuous pressure – starving the empire of fuel and other critical supplies – wore down Japan's ability to sustain war regardless of Japan's battlefield performance. War is waged on supply chains as much as battlefields, and the strategic economic isolation of Japan by 1945 due to these Pacific patrols devastated Japanese supply lines, contributing to its economic and military collapse.

Decentralized Command – Trusting Subordinates

Some have argued that Nimitz's greatest legacy was not in tactics but in leadership. He was self-assured yet humble, and welcomed input from junior officers. He selected discerning staff officers, sought the advice of tactical commanders, and encouraged open debate and the sharing of ideas. Historian E.B. Potter observed, "Nimitz was a master at making good men better."

This command style fostered innovation and initiative among his subordinates. Nimitz also delegated operational decisions, once again fostering initiative. This environment ensured rapid decision-making in dynamic battles, and allowed the U.S. Navy to adapt quickly to the changing realities of the Pacific War. Nimitz basically provided strategic intent but trusted commanders with tactical execution. He was not a tactical commander in every engagement, but delegated effectively to brilliant subordinates.

Nimitz avoided micromanagement, and his calm leadership style empowered skilled subordinates like Raymond Spruance, William Halsey and Marc Mitscher, trusting them to exploit opportunities and adapt to battle conditions while interpreting his strategic intent in a fluid, chaotic battlespace. This included Spruance's leadership at Philippine Sea, and Halsey's aggressiveness at Leyte Gulf.

This balance of Nimitz's centralized strategic planning with decentralized execution enabled the U.S. Navy to operate across 6,000 miles of ocean while maintaining cohesion and initiative.

Nimitz's trust in subordinates prevented paralysis during fast-changing environments by allowing flexibility in vast operations, as seen in Spruance's and Halsey's contrasting styles. His calm demeanor was able to unify diverse personalities and forces, and his trust in subordinates, which was evident at Leyte Gulf – the largest naval battle in history with over 200,000 personnel involved – ensured success despite operational challenges.

Nimitz's Enduring Legacy

Chester W. Nimitz's embracing of carrier warfare as the primary naval weapon over (outdated) battleship doctrine, practice of turning codebreaking into battlefield wins, island-hopping strategy to bypass heavily fortified islands and conserve resources, submarine campaigns to cripple Japan's economy by sinking its merchant fleet, and centralized strategy while employing a decentralized command of trusting skilled subordinates, won battles and reshaped war at sea. Nimitz's tactics were revolutionary because they integrated intelligence, carrier warfare, and strategic restraint, shifting naval doctrine from battleship-centric engagements to flexible, air-driven operations.

Under Nimitz's command, the U.S. Navy achieved a decisive strategic advantage after Midway and never relinquished it. The campaigns he directed – from the defensive stand at Coral Sea to the outright victories at Midway, the Philippine Sea, and Leyte Gulf – secured Allied control of the Pacific, ending Japan's naval power. The carrier task force model he perfected became the blueprint for every major navy in the world for the remainder of the century.

Nimitz's legacy is perhaps best summed up by the words of Admiral Ernest King: "Nimitz never lost sight of the main objective, never lost his nerve, and never lost his faith in the men under his command."

CHAPTER 10:
ANDREW BROWNE CUNNINGHAM (1883-1963)

Admiral Andrew Browne Cunningham, known as "ABC," was a cornerstone of British naval success in the Mediterranean during World War II, leading the Royal Navy's Mediterranean Fleet to victory against a numerically superior Italian navy. Cunningham is widely regarded as the greatest British admiral since Nelson, and transformed the Royal Navy's Mediterranean Fleet into a formidable force that won a series of crushing victories, crippled the Italian battle line at Taranto (1940), and maintained critical supply lines.

Unlike cautious contemporaries, Cunningham pursued enemy fleets regardless of odds or conditions, including Luftwaffe attacks. His aggressive tactics, innovative use of air power, and unwavering commitment to logistics secured critical sea lanes and crippled Axis operations by securing victories at Taranto, Cape Matapan (1941), and in the Malta convoys. As First Sea Lord (from October 1943) he saw the naval side of Overlord, the largest amphibious operation in history.

Cunningham's refusal to yield initiative, even when outnumbered, forged British dominance in the Mediterranean during World War II. He led the Royal Navy's Mediterranean Fleet through a gauntlet of challenges – Italian surface fleets, Luftwaffe air power, and Axis convoys – and emerged time and again victorious. His warfare philosophy was simple: attack always. His famous statement, "It takes three years to build a ship; it takes three centuries to build a tradition," reflects his dedication to the Royal Navy's offensive ethos.

Aggressive Pursuit

Cunningham's all-out offensive posture was encapsulated in his order to "sink, burn, and destroy – let nothing pass." He prioritized relentless offense even when facing the bad odds of fighting equal or superior forces, and often in unfavorable conditions such as at nighttime or in poor visibility. This Nelsonian approach sunk Italian forces and enabled the British to establish a psychological hold over the Mediterranean.

Cunningham's calculated risks and his trust in his crews' training set him apart from most commanders. British crews, rigorously trained in gunnery and damage control, often outperformed their Italian counterparts in gunnery duels and damage control, a critical edge in close engagements. N.A.M. Rodger, for instance, estimates that British ships fired 20% faster than Italian vessels (*The Command of the Ocean*). Furthermore, Cunningham engaged despite any Luftwaffe threats, which forced decisive outcomes. At the Battle of Crete (May 1941), Cunningham evacuated 16,000 troops under Luftwaffe bombardment, saving Allied forces.

Cunningham pursued enemy fleets relentlessly, and refused to let the enemy regroup by maintaining constant pressure. At the Battle of Calabria (July 1940), despite numerical disadvantages he pressed the attack and pursued the Italian fleet with *HMS Warspite*, securing a lucky, extreme-range hit on the Italian flagship *Giulio Cesare* despite being outnumbered. This hit caused minor damage but had a major psychological impact in causing the Italian fleet to disengage.

His unrelenting offensive mindset kept Axis forces on the defensive. It was also seen at the Battle of Cape Matapan where after detecting the Italian fleet through Ultra codebreaking (which enabled preemptive strikes), he chased Admiral Iachino's fleet down at night, sinking three heavy cruisers (*Zara*, *Fiume*, *Pola*) and two destroyers.

Night-Fighting Mastery

Under Cunningham, the Royal Navy excelled at night combat, and exploited this superiority by seeking night engagements using radar for precise targeting, a capability the Italians lacked who were also deficient in night-fighting experience.

British ships equipped with Type 279/284 radar could achieve precise targeting in darkness. Combined with training for blind-fire gunnery, this transformed darkness into a tactical advantage for the British, allowing their forces to strike with precision and surprise. Naval historian Vincent P. O'Hara *(Struggle for the Middle Sea)* notes that Italian commanders feared night engagements whereas Cunningham sought them.

The night fighting mastery at Cape Matapan involved radar-directed fire from the *Warspite, Valiant*, and *Barham* in pitch-black conditions. The British unleashed crushing broadsides into Italian ships and annihilated the cruisers, with 2,400 sailors lost. Additionally, at the Battle of the Duisburg Convoy (November 1941), Cunningham also used radar to sink an entire Italian convoy.

Cunningham speed and secrecy was amplified by his integration of Ultra intelligence into naval operations. He would use intelligence reports to position the British fleet for maximum effect, guiding ambushes and ensuring surprise. Cunningham aptly noted, "Good intelligence is worth a dozen battleships."

Carrier-Based Integration of Naval Air Power

Cunningham's embrace of carrier aviation was groundbreaking. His authorization of the Taranto raid launched the first all-air naval attack in history, a daring gamble that changed naval warfare. On 11–12 November 1940, under his command, 21 Swordfish biplanes from the carrier *HMS Illustrious* executed the first all-air strike on a battle fleet resting in port (at Taranto) — a surprise attack that crippled Italy's battlefleet. The attack proved that traditional battleships could be neutralized by air strikes alone.

The raid sank the *Conte di Cavour* and damaged the *Littorio* and *Caio Duilio*, which halved Italy's capital ship strength. The Taranto attack marked a turning point in naval warfare – it proved that carrier-based air power could neutralize battleships, thus influencing global naval strategies. The all-air attack on a port also foreshadowed the Japanese attack on Pearl Harbor.

A Logistical Focus

Cunningham knew that logistics won wars; supply lines sustain operations so they underpin victory. His strategic focus on logistics, particularly defending the island of Malta through convoy operations, was critical to the Allied success in North Africa.

"Malta must be held at all costs," he said, directing destroyers and cruisers to proactively hunt U-boats and aircraft to reduce convoy losses to the island. Feints, such as false convoys, also diverted Axis forces. Operation Pedestal (August 1942) delivered vital supplies to Malta despite losing the aircraft carrier *Eagle* and nine merchantmen. Knowing that defending Malta ensured Allied control of the Mediterranean, Cunningham's willingness to endure such heavy losses for strategic gains underscored the importance of logistics in warfare.

In the Eastern Mediterranean Operations (1941–42), Cunningham used *HMS Formidable* to relentlessly harass Axis supply convoys, which primarily carried troops, fuel, vehicles, and ammunition from Italy to North Africa to support Rommel's Afrikakorps and Italian forces in Libya. By early 1941, over 90% of Axis supplies to North Africa traveled by sea from Italy to Tripoli/Benghazi. Cunningham therefore aimed to strangle this supply line through Malta-based air and naval strikes, surface actions, and carrier operations.

Essentially, Cunningham's merciless focus on supply lines, despite heavy losses, strangled Axis supplies to North Africa (Rommel's logistics) and ensured the Allied victory in North Africa, vindicating his offensive ethos.

Psychological Warfare

Cunningham cultivated a reputation for invincibility (dubbed the "Cunningham Effect") in order to boost British morale and inspire his men to fight beyond normal limits. He also hoped to demoralize Italian commanders who might then avoid direct engagements.

The so-called "Cunningham Effect" definitely paralyzed Italian naval actions because after Matapan, Italian Admiral Iachino refused to leave port, fearing Cunningham's fleet. Post-Matapan (1941), the *Regia Marina* avoided major fleet actions, which ceded sea control to British dominance. In Operation Harpoon/Vigorous (June 1942) the Italian hesitancy due to fear of Cunningham enabled British convoys to reach Malta.

Cunningham's reputation for invincibility stood as a shadow looming over every Italian sortie and his leadership galvanized his fleet, enabling them to endure the sometimes deadly Luftwaffe attacks.

Why Cunningham's Tactics Surpassed His Peers

Unlike his Axis counterparts, constrained by rigid orders, Cunningham granted captains flexibility, echoing Nelson's practice of decentralized command. He integrated radar and aviation, emphasized logistics (contrasted with Axis neglect of supply lines), exploited intelligence, and weaponized psychology. He echoed Alfred Thayer Mahan's belief in decisive engagement, yet adapted it to the air and radar age.

While peers like Japan's Admiral Yamamoto relied heavily on carriers, Cunningham achieved a balanced fusion of air, surface, and subsurface forces. He blended traditional naval warfare methods with the new modalities of modern technology. His tactics of relentless engagements, night-fighting superiority, air power integration, logistical tenacity, intelligence exploitation, and psychological warfare devastated the Axis and secured the Mediterranean.

CHAPTER 11:
DAVID FARRAGUT (1801-1870)

The U.S. Navy's first full admiral, David Farragut, showcased aggressive leadership that led to crucial Union victories in riverine and coastal battles during the American Civil War, most notably at the Battle of Mobile Bay (1864) and at the capture of New Orleans (1862).

His signature tactics combined traditional naval principles with innovative adaptations. He turned riverine warfare into a precise instrument, treated fortified defenses as obstacles to bypass rather than besiege, and visibly led from the front, all of which enabled him to win against formidable odds and cemented his legacy as a tactical genius.

Farragut excelled at forcing engagements and pressing his advantage, and remained undefeated in significant naval battles. Naval historian Alfred Thayer Mahan compared Farragut to Horatio Nelson in his willingness to take risks, and one might summarize his naval career as one of aggressive initiative.

When Farragut assumed command of the West Gulf Blockading Squadron in 1862, naval warfare was dominated by siege mentality – the belief that heavily fortified positions required systematic softening through prolonged artillery bombardment before fleet operations could proceed. As naval historian Alfred Thayer Mahan observed in his biography of Farragut, "The flag-officer, having heard all opinions expressed by the different commanders, is of the opinion that whatever is to be done will have to be done quickly." This philosophy of speedy aggression would become the cornerstone of Farragut's warfare approach.

Other admirals of Farragut's era typically remained bound by conventional doctrine and cautious approaches, but Farragut's aggressive approach was rooted in doing the homework of meticulous preparation (gathering detailed information on enemy defenses, river conditions, and weather), and in his belief that hesitation in battle was more dangerous than bold action. Unlike cautious naval commanders who preferred long bombardments, Farragut often risked close-quarters engagements to achieve swift victories.

He revolutionized naval warfare through a combination of audacious tactics (such as "running-the-gauntlet" and combined fleet operations) along with an unprecedented willingness to accept calculated risks for decisive results. His analytical approach of understanding both his own fleet's capabilities and enemy weaknesses represented a departure from the rigid tactical thinking that had dominated naval warfare since the Age of Sail.

Aggressive Offensive Action

Farragut's strategic philosophy centered on strong leadership combined with rapid execution. Rather than engaging in prolonged operations that allowed enemy reinforcement and preparation, Farragut typically chose immediate action when circumstances held the potential of decisive results.

At the capture of New Orleans, for instance, his decision to run past the forts under heavy fire after a failed bombardment, rather than retreat or attempt a prolonged siege, exemplified his philosophy of aggressive forward action. Despite fierce resistance, the Union ships broke through, destroyed Confederate naval forces, and reached New Orleans. Farragut had attacked rather than become paralyzed by tactical complexity.

Farragut's signature tactic was aggressive initiative (aggressive offensive maneuvers), such as his bold decisions to charge through enemy minefields and fort bombardments at full speed to seize objectives or to engage at close range, which usually left Confederates in complete surprise.

Unlike his predecessors, who often relied on cautious blockades or static bombardments, Farragut reasoned out the limitations of land fortifications, which remained dependent upon slower-firing artillery systems, against the development of steam powered wooden ships equipped with improved artillery. These developments had created new tactical scenarios where there was a clear imbalance between the offense and defense. He developed tactics that exploited the weaknesses of fortifications, leveraging the speed and mobility of his steam-powered ships to force their way past any land-based defenses.

His success at doing so challenged the accepted wisdom of the time that "forts commanded their adjacent surrounding waterways." His offensive aggressiveness was made possible only because steam power's ability to maintain speed and maneuverability was independent of wind conditions, making rapid passage past forts feasible where sailing ships would have been vulnerable. Steam-powered ships could now navigate rivers and bays where sailing ships struggled, outpacing any Confederate responses.

The use of steam therefore constituted a leap forward in naval possibilities, allowing unprecedented speed and control that enabled Farragut to outmaneuver slower sailing vessels and also minimize any exposure to shore-based artillery. By avoiding prolonged artillery bombardment from forts and shore batteries, Farragut preserved his fleet's strength and could instead focus on more strategic objectives, such as controlling trade routes or isolating enemy forces.

As an example of his aggressiveness, Farragut's fleet of 17 ships ran a gauntlet past Forts Jackson and St. Philip, navigating their 115-gun barrage and fire rafts at night. Despite taking heavy fire from the Confederate forts and a melee with defending Confederate vessels, his aggressive charge up the Mississippi River, running-the-gauntlet by using steam to maintain full speed under heavy fire, overwhelmed Confederate defenses, and captured the key Confederate port of New Orleans with minimal losses (1 ship sunk). This deprived the Confederacy of its largest city and enabled further Union advances up the Mississippi River. His success clearly demonstrated that Confederate riverine defenses, however impressive, could be overcome through audacious execution.

Farragut's run-the-gauntlet strategy enabled him to regularly bypass Confederate forts and mines, which were called "torpedoes" at the time, rather than engage in prolonged sieges. It exploited the Union's naval superiority in mobility whereas Confederate forces relied on fixed positions and had no time to adjust fire because of his speed. His success also shocked Southerners because Confederate gunners were usually unprepared for such audacity. The speed of Farragut's fleet and his night attacks minimized Union losses, demonstrating that will and aggressive momentum could render even powerful forts less important.

At the Battle of Mobile Bay, Farragut faced the concentrated fire of Fort Morgan's 46 guns, a dense minefield, and the powerful ironclad *CSS Tennessee*, but Farragut's 18 ships charged through torpedoes (mines) with his famous cry, "Damn the torpedoes! Full speed ahead!" His rapid advance neutralized the minefield's threat and allowed his fleet to engage at close range, sinking the Confederate fleet (1 vs. 4 ships lost) and capturing the bay to secure a vital Union blockade point.

The rapid movements of Farragut's fleet denied the Confederate artillery time to adjust fire, and exploited the unpreparedness of Confederate gunners who did not expect the audacity of his close-range assaults.

In June 1862, Farragut's fleet ran past Vicksburg's formidable batteries, using speed to once again minimize any damage from fort bombardments, to link with Union forces upstream, an early attempt to pressure the city. His dashing maneuvers disrupted the Confederate supply lines, contributing to the city's eventual fall, and became a new standard option for naval support.

Bypassing Fortified Defenses by Defying Conventional Wisdom

As explained, Farragut's strategy of bypassing heavily fortified coastal defenses – charging past them at maximum speed with guns blazing rather than engaging in prolonged siege operations – was a radical departure from traditional naval doctrine, which emphasized bombarding forts into submission. Farragut never accepted the assumed limitations imposed by enemy fortifications. Influenced by his mentor, Commodore David Porter, Farragut believed that forts were secondary to controlling waterways.

By avoiding prolonged engagements with forts, Farragut preserved his fleet's strength and focused on decisive objectives, such as controlling trade routes or isolating enemy forces. His steam-powered ships could navigate rivers and bays where sailing ships struggled, outpacing Confederate responses and allowing access where none was previously possible.

Hence, at New Orleans, Farragut bypassed Forts Jackson and St. Philip, ignoring their 115 guns to capture the key Confederate port. At Mobile Bay, Farragut sailed past Fort Morgan's land-based guns and a minefield, focusing on the Confederate fleet and capturing the bay. His tactic neutralized shore defenses and secured a vital Union blockade point. At Port Hudson (1863), Farragut's fleet ran past 20 Confederate guns to establish a blockade, cutting off Confederate supplies. Though costly (2 ships damaged), his bypass ensured Union control of the Mississippi River.

Adaptive Fleet Formation and Lashing Tactics

Farragut also developed innovative fleet formations (instead of rigid line-ahead) adapted to specific tactical challenges. At Mobile Bay, he faced a narrow, mine-infested channel with shore forts on both sides, and the Confederate ironclad *Tennessee* plus three gunboats waiting inside.

His 14 wooden steam frigates and 4 ironclads steamed in pairs (a wooden ship was lashed to the port side of an ironclad for protection), forming two parallel columns to pass Fort Morgan's guns. This repeated a tactic that he had used earlier at the Battle of Port Hudson.

This innovation served multiple purposes. It provided mutual support for damaged vessels by sharing propulsion because if the wooden ship lost steering or power then the ironclad could tow or push it, and if the ironclad grounded the sloop could pull it off. Guns on both sides could fire simultaneously, and lashing the ships together enabled them to maintain speed and formation integrity under fire. Most of all, this protected vulnerable wooden hulls behind ironclads.

Farragut's lashed-pair, adaptive column tactic became the U.S. Navy textbook method for forcing fortified narrows through the Spanish American War and was still taught at Annapolis into the 1920s.

Combined Operations Integrating Naval and Land Forces

Farragut excelled at coordinating naval operations with army maneuvers – such as troop landings, bombardments and blockages – a tactic less common among his peers who often operated independently. His riverine operations integrated steamships and gunboats with army support to extend naval power into inland waterways, which was a critical innovation in the Civil War. Farragut demonstrated through his riverine campaigns that naval power could be decisive in a continental war, not by fleet battles alone, but by enabling and sustaining the army's campaign deep inland.

Farragut clearly demonstrated that naval power was most effective when it integrated, in synchronized fashion, with land operations. During the Siege of Vicksburg, Farragut's fleet supported General Ulysses S. Grant's land campaign by blockading and bombarding the city of Vicksburg's defenses, which enabled Grant's army to besiege the city. His naval support cut Confederate supply lines and forced Confederate garrisons to split their focus between land and river defenses, which were pivotal to the campaign's success.

At New Orleans, Farragut coordinated his fleet passage with General Benjamin Butler's troop landings, ensuring that the troops from Union transports could land under the protection of Farragut's batteries. Both the fort and city were captured after Farragut's ships silenced the Confederate defenses, which was a coordination that secured the mouth of the Mississippi. Again, Farragut didn't simply coordinate with land forces but created synergistic operations where naval and military elements mutually supported each other's objectives.

At the Battle of Memphis, Farragut's gunboats defeated the Confederate River Defense Fleet (8 vs. 8 ships), securing Memphis for Union forces. His naval coordination with on-ground army units ensured rapid occupation of the city.

Farragut's practice of using gunboats for riverine warfare was ahead of its time, providing an early model for integrated riverine and coastal warfare. His integration of naval and land forces also represented an evolution beyond traditional naval thinking. Because both forces were needed, he admiral didn't simply coordinate with land forces but made sure that naval and military elements mutually supported each other's objectives. As historian Christopher L. Kolakowski notes, Farragut "needed ironclads to pass Fort Gaines and Fort Morgan, and Army troops to assist in capturing the forts themselves."

Deceptive Feints and Night Movements

Farragut's use of deception was also rare in naval warfare at the time. Most commanders of his day relied on brute force, but Farragut was adept at employing trickery in warfare. He took a page from Sun Tzu that "all warfare is based on deception," and used feints to force Confederate commanders to waste ammunition that revealed their gun positions.

For instance, at New Orleans Farragut used false bombardments to distract Confederate gunners before his main fleet ran past, and at Mobile Bay he ordered nighttime mine-clearing operations to deceive Confederate troops about his fleet's movements.

When the monitor *USS Tecumseh* struck a mine and sank, the leading wooden ship *Brooklyn* stopped in alarm, and the whole line of ships drifted in confusion while under the very guns of Fort Morgan. At this critical juncture, Farragut's decisive command – "Damn the torpedoes! Full speed ahead!" – broke the psychological deadlock, overcame fleet paralysis and restored offensive momentum.

Visible Courage

Farragut's leadership under fire was a critical, often overlooked tactic that galvanized his crews during high-risk maneuvers. Farragut's personal bravery – such as being lashed completely calm and completely fearless to the rigging of *USS Hartford* at Mobile Bay (the hottest spot), and cutting the lashings of a fire-raft with his own hands when the *USS Hartford* ran aground directly under the muzzles of Fort St Philip – not only boosted morale and inspired his men to fight harder but ensured that crews followed orders precisely.

His visible courage raised his crew's spirits during high-risk maneuvers and inspired them to execute daring tactics with resolve, such as charging minefields or forts. His personal leadership ensured the crew's cohesion during chaotic battles and turned potential routs into victories.

The Farragut Legacy

Farragut's famous cry at Mobile Bay – "Damn the torpedoes (mines)! Full speed ahead!" – encapsulates his philosophy of decisive action. It was less a traditional naval command and more a Napoleonic imperative: bold action over cautious calculation. His tactics of aggressively charging with speed, bypassing forts to seize objectives, integrating naval and army forces, deception, and leading visibly to inspire crews, set him apart from contemporaries who still adhered to slow, methodical naval warfare. He established American naval principles that would influence naval strategy well into the modern era.

Farragut's pioneering tactics transformed brown-water naval warfare because they harnessed steam technology, and enabled the shift from sail-era caution to aggressive, dynamic, offensive operations. His use of steam to force his way past fortified coastal defenses was a giant leap forward in naval tactics. His riverine operations also pioneered joint army-navy campaigns, thus becoming a model for modern amphibious warfare, and his aggressive maneuvering and fort-bypassing strategies proved that controlling waterways could win wars without engaging every fort.

In addition to his revolutionary tactics, Farragut's win-loss record (~3 major wins, no losses in pitched battles) reflects his ability to overcome numerical and technological disadvantages, such as Confederate ironclads and mines. His victories at New Orleans, Mobile Bay, and Vicksburg secured Union control of the Mississippi and Gulf, splitting the Confederacy and ensuring Union triumph in the American Civil War.

CHAPTER 12:
HAYREDDIN BARBAROSSA (c. 1478–1546)

Admiral Hayreddin Barbarossa (c. 1478–1546), appointed Kapudan Pasha (Grand Admiral) of the Ottoman Empire by Suleiman the Magnificent, dominated the Mediterranean Sea in the 16th century, and secured Ottoman naval supremacy through a series of decisive victories against numerically superior Christian fleets. He was originally an Ottoman corsair (privateer or pirate) before later becoming an admiral.

Known as the "Red Beard" for his fearsome appearance, Barbarossa's innovative tactics – *guerre de course* (commerce raiding), decentralized fleet operations, coastal warfare, integration of gunpowder artillery, feigned retreats and ambushes, effective intelligence networks, and psychological dominance – set him apart from many predecessors and his peers. Hayreddin Barbarossa's strategies did not just win battles; they redefined the rules of naval power in the Mediterranean and transformed it into an Ottoman lake.

Barbarossa actively cultivated an image of invincibility. Many ports capitulated without a fight, their defenders frightened because of his reputation. These cities agreed to pay tribute, directly extending Ottoman influence while adding to Ottoman coffers. Barbarossa's reputation caused such fear that it often routed opponents before combat started, and was instrumental in sometimes causing enemy fleets to forgo battle.

Barbarossa's record – undefeated in major engagements, architect of strategic alliances, and pioneer of artillery-based galley fleets – places him among history's greatest naval commanders.

Guerre de Course (Commerce Raiding)

Barbarossa pioneered large-scale commerce raiding (*guerre de course*) with the intent of disrupting trade to weaken adversaries economically and psychologically. He targeted Spanish, Venetian, and Genoese merchant vessels to cripple their economies and sow fear across Christian Europe.

Commerce raiding starved Christian adversaries of revenue, forcing them to divert ships to convoy protection, thus reducing their offensive capabilities. The psychological impact – fear of sudden raids – prompted coastal cities to pay tribute, further enriching Ottoman coffers.

By targeting Spanish and Venetian merchant fleets, Barbarossa thereby drained national resources and strained enemy logistics. While Christian admirals aimed for decisive Trafalgar-like clashes, Barbarossa focused on a slower, deeper victory: disrupting the trade routes to strangle the trade that funded his enemies' navies in the first place. This tactic was a new development in the galley-dominated 16th century where fleets typically clashed in pitched battles, and it shifted the military focus to long-term economic attrition.

Barbarossa targeted Italian trade routes in the 1520s–1530s, sinking or capturing dozens of merchant vessels and breaking the Christian supply chain. This disrupted the economies of Venice and Genoa, weakening their naval alliances. Another example is that he targeted coastal raids near Sicily and Naples (1534–37) to also limit their naval resources.

Barbarossa's corsairs plundered Spanish merchant ships and coastal towns, capturing thousands of slaves and goods. In the Balearic Islands Raids (1535), Barbarossa's fleet ravaged Majorca and Minorca, capturing 6,000 prisoners – thousands of merchants and Spanish troops – which forced Spain to divert resources from its fleet to fund massive coastal fortifications.

Decentralized Fleets Tactics: Agile and Adaptive Operations

Barbarossa's fleets operated in decentralized squadrons, which enabled agile, adaptive operations that could swiftly harass or envelop enemies, thus making his fleets unpredictable. This pioneering approach created fleet adaptability in the chaotic Mediterranean where weather and geography demanded fluid tactics.

This fluidity stood in stark contrast to the rigid, line-astern formations of adversaries like Andrea Doria. It allowed Barbarossa's galleys to strike independently, regroup swiftly, or scatter to avoid encirclement, drawing inspiration from earlier Muslim naval tactics. Barbarossa's decentralized squadrons exploited gaps in enemy formations, targeting isolated ships and avoiding large-scale engagements unless it was advantageous. This flexibility countered the Holy League's numerical superiority, and in *The Influence of Sea Power Upon History* Mahan compared this to "Carthaginian flexibility against Roman rigidity."

At the Battle of Preveza (1538), for instance, Barbarossa's fleet, slightly outnumbered by the Holy League's combined force, used small, decentralized squadrons to harass and fragment Andrea Doria's formation, sinking or capturing several ships with minimal losses.

In the Algiers Defense (1541), while facing Charles V's 500-ship armada, Barbarossa's 80 galleys used decentralized attacks to disrupt landings, which forced a Spanish retreat.

The coordinated tactics he pioneered were successfully employed by his successor, Dragut, in the capture of Tripoli in 1551.

Feigned Retreats and Ambushes

Barbarossa frequently employed feigned retreats to lure enemy fleets into vulnerable positions, such as shallow waters, and then his galleys would counterattack with hidden reserves or ambushes. His disciplined crews maintained order during retreats, unlike pursuing enemies who often broke formation.

This tactic, inspired by Muslim predecessors like Tariq ibn Ziyad, exploited the overconfidence of Christian commanders. Barbarossa's feigned retreats defied convention in an era where direct galley clashes dominated because they prioritized the use of psychological manipulation over brute force.

As an example, during the Calabria Raids of the 1520s, Barbarossa lured Spanish galleys into shallow coastal waters with feigned retreats, then ambushed them with hidden squadrons, sinking several ships. At Preveza, Barbarossa's feigned withdrawal baited Doria's fleet into a disorganized pursuit, allowing his galleys to regroup and strike to sink many of Doria's ships.

At Djerba in 1560, Ottoman forces under Piyale Pasha, employing Barbarossa's strategic playbook, used a feigned retreat to lure the Spanish fleet into a trap, resulting in a catastrophic defeat for the Holy League.

Mastery of Hit-and-Run Coastal Warfare & Galley Tactics

Barbarossa avoided large-scale fleet battles unless they were strategically necessary. He favored hit-and-run coastal raids that leveraged his fleet's superior maneuverability and his intimate knowledge of Mediterranean geography. For instance, he leveraged his coastal knowledge of Tunis (1534) for hit-and-run strikes, outmaneuvering Spanish defenders, and landing troops under cover of night so as to capture the city with minimal naval losses.

At Algiers, Barbarossa's fleet raided Spanish outposts, using shallow waters to evade pursuit, and he strategically secured Algiers as an Ottoman base. At Tripoli (1551), his fleet exploited coastal currents to launch a surprise attack, capturing the city from the Knights of Malta.

These coastal waters tactics, rooted in his corsair experience, allowed rapid strikes and withdrawals, which frustrated larger Christian fleets. His successful coastal warfare extended Barbarossa's operational range by using North African ports for resupply and refuge. Essentially, his raids disrupted enemy logistics while avoiding decisive engagements, thereby preserving his fleet.

Integration of Artillery onto Galleys

Barbarossa led yet another naval transformation by integrating artillery into galley warfare, pioneering the use of ship-mounted cannons on galleys that transformed them into floating fortresses. Cannons reduced the necessity for risky boarding actions, so this shifted naval operations from boarding-centric warfare to standoff artillery engagements.

This new naval strategy was revolutionary in the 16th century, as Christian fleets still prioritized ramming and boarding. His galleys' cannons aimed to shred mast rigging and smash hulls at distance, crippling the enemy's mobility and morale long before the boarding pikes came out.

At the Battle of Preveza, his ship artillery outgunned Doria's fleet and softened the enemy longships before close engagement. Barbarossa's galleys used concentrated cannon fire to disrupt the Holy League's formation and sink 10 ships as stated. At the Siege of Algiers, Barbarossa's cannon-armed galleys and their effective shoreline bombardment repelled Spanish assaults, sinking 12 ships and forcing a retreat. At Tunis, his fleet's artillery bombarded Spanish defenses, enabling a swift landing and capture.

Intelligence Networks: The Eyes of the Fleet

Barbarossa's extensive spy network, a type of tactics often overlooked, provided critical intelligence on enemy movements, fleet compositions, and coastal defenses. His corsair background fostered connections with local informants across the Mediterranean – essentially a network of North African and Levantine spies – giving Barbarossa a strategic edge of accurate intelligence allowing him to strike at vulnerable targets, avoid traps, and exploit enemy disorganization.

Because his spy network shaped nearly every movement, naval historian John B. Hattendorf observes, "Barbarossa's victories began with his spies" *(Oxford Encyclopedia of Maritime History).* At Preveza, spies informed Barbarossa of Doria's fleet deployment, enabling his decentralized attacks. At Tunis, informants revealed Spanish garrison weaknesses, allowing a surprise coastal assault. At Djerba, Barbarossa's strategy, executed by his successor, relied on intelligence about Spanish fleet movements, leading to a decisive ambush.

Much of his success was due to superior naval intelligence, which was far better than that of his Christian opponents.

Leadership and Morale: The Barbarossa Legend

Barbarossa's fearsome reputation as a corsair-turned-admiral was a deliberate tactic for cultivating the psychological expectation of dominance that often won battles before they began. As we have seen with several other great admirals, his strong reputation intimidated enemies, causing some to flee or surrender, while his charismatic leadership ensured the discipline of his crews during complex maneuvers.

As examples, at Tripoli his legend prompted Maltese defenders to hesitate, easing the city's capture, and at Preveza, Barbarossa's reputation frightened some Holy League ships to flee before engagement.

Barbarossa's leadership inspired loyalty among diverse crews, from Ottoman sailors to North African corsairs. At Algiers, his leadership rallied crews to repel a larger Spanish armada, and he was able to maintain cohesion under heavy fire. Barbarossa's personal charisma was the glue that held his polyglot, corsair fleet together, transforming them from raiders into a disciplined navy.

Impact and Lasting Legacy

Barbarossa's tactics were revolutionary because they combined economic warfare, tactical flexibility, and the early adoption of gunpowder artillery, which all challenged the European practices of rigid galley formations and boarding tactics. His *guerre de course* anticipated modern commerce raiding, as seen in later figures like Francis Drake, while his decentralized squadrons foreshadowed 20th-century distributed operations, per Wayne Hughes *(Fleet Tactics)*. His use of intelligence and psychological dominance also set a standard for asymmetric warfare, influencing admirals like Horatio Nelson and Dragut.

Barbarossa's win-loss record (~5 major wins, 1 minor loss) reflects his ability to defeat larger fleets through superior naval tactics and intelligence. The later heroic chronicle *Gazavat-ı Hayreddin Paşa* (*Conquests of Hayreddin Pasha*), details these naval campaigns, cementing his legendary status. His victories secured Ottoman naval supremacy in the Mediterranean for decades, reshaping its trade routes and power dynamics to favor Istanbul. In the centuries following his death, no fleet would clear the Serai Point of Istanbul, where he was buried, without firing a salute at his mausoleum. This practice disappeared during the Tanzimat period of Ottoman history but was revived by the Turkish Navy in 2019.

Historian Andrew Lambert sums up his legacy: "Barbarossa turned piracy into strategy, making the Ottoman navy a global power" *(Admirals)*. His influence persisted in Ottoman naval doctrine, with successors like Turgut Reis adopting his tactics.

CHAPTER 13:
FRANCIS DRAKE (c. 1540-1596)

Admiral Sir Francis Drake (c. 1540-1596), celebrated English vice-admiral, was a formidable naval commander in the Elizabethan era who disrupted Spanish naval dominance and secured England's rise as a maritime power. Drake's repertoire of tactics was revolutionary because it discarded the rulebook of traditional fleet doctrines and formal naval engagements to fight a new type of war. His signature tactics – commerce raiding and privateering, rapid strike maneuvers, fireship attacks, coastal raids and amphibious operations – were as varied as effective. They set him apart from predecessors like Don Álvaro de Bazán, who favored traditional galley battles.

Drake's overarching strategy was born from a clear-eyed understanding of England's geopolitical position in the world and the vulnerabilities of its Spanish adversary. Drake recognized that England's security and prosperity both depended on breaking Spain's monopoly over the seas and the New World, and his strategies were developed with this in mind.

His campaigns were therefore designed not merely as raids but as blows to the very infrastructure of Spanish imperial power, forcing Spain to divert resources and to rethink its global strategy. Drake's strategies included a blend of piracy-inspired tactics, strategic deception, and the exploitation of the mobility of sail-powered ships to challenge the rigid fleet formations of the Spanish.

Coastal Raids and Privateering

Drake's hallmark tactic, operating under Queen Elizabeth I's letters of marque, was commerce raiding and privateering. He targeted Spanish treasure fleets and colonies with raids in order to cripple Spain's naval power and economy. Drake's coastal raids and amphibious operations targeted Spanish ports and colonies, combining naval strikes with land assaults to maximize disruption.

Unlike predecessors who focused solely on fleet battles at sea, Drake integrated naval and ground forces, foreshadowing modern amphibious warfare. He combined his naval bombardments with amphibious assaults at the capture of Santo Domingo (1586) and the Cartagena Expedition (1586). Drake's assault on Cartagena followed his typical pattern: a naval bombardment and amphibious landing (usually landing troops covertly at night) to seize the city, followed by a ransom demand as a tool for both profit and psychological warfare.

Drake would specifically conduct raids against Spain's naval infrastructure and convoys that isolated and weakened Spain's maritime defense. His raids turned piracy into a strategic weapon because they prioritized captured gold, silver, and supplies that weakened Spain's war machine, thus destabilizing the Spanish Empire due to the loss of revenue. His coastal raids disrupted Spanish logistics and morale, while amphibious operations captured key strongholds and weakened Spain's colonial empire.

This was a revolutionary shift in objective for the 16th century. Where others sought to control land, Drake sought to bankrupt an empire – elevating commerce raiding from a criminal act to a strategic doctrine. For instance, in a strategic move at Cadiz (1587) Drake's ships and land assault attacked the Cadiz harbor, destroying Spanish vessels and associated supplies that then delayed the Spanish Armada's launch against England. This preemptive raid is often cited by naval strategists as a classic example of offensive action disrupting an adversary's invasion plans.

Drake's basic strategy was elemental: bleed the enemy's economy to death, delay its naval operations and sow fear to weaken its resolve. By targeting Spain's treasure ships and colonies, Drake consistently drained the Spanish resources needed for its fleet's maintenance, forcing Spain into costly defensive measures. His raids also sowed panic, and diverted Spanish ships from offensive operations. Drake's raids inflicted direct financial costs on Spain and this relentless hemorrhage began to tilt the Atlantic's economic axis. Traditional admirals were seeking decisive fleet battles, but Drake's hit-and-run raids were far more effective than the conventional line-of-battle approach because they were effectively crippling enemy logistics.

In his early raids (1572-73), Drake's small fleet targeted Panama's treasure port of Nombre de Dios, capturing silver and gold and disrupting Spain's colonial economy. At San Juan de Ulúa (1568), Drake with John Hawkins attacked Spanish treasure ships, capturing loot despite a larger Spanish response. At Santo Domingo, Drake's fleet and troops captured the city, looting supplies and ransoming it, which disrupted Spain's control of the Caribbean.

Rapid Strike Maneuvers

Drake exploited the weather gauge (wind advantage) to dictate his engagements because he relied on maneuverability, unlike Spanish admirals who relied on large, lumbering galleons. His own predecessors had relied on slow, heavily armed ships but Drake used lighter, faster, more maneuverable vessels to launch surprise attacks and evade pursuit. Drake's tactics were revolutionary for exploiting the potential of sail technology. His smaller ships avoided direct engagements with superior forces, ensuring tactical initiative.

Drake's rapid strike maneuvers, leveraging the speed and maneuverability of smaller sail-powered ships, allowed him to outpace and outmaneuver larger Spanish fleets that were largely cumbersome galleon formations. Through rapid strikes Drake could hit and withdraw before his enemies could respond. Because his ships were superior in tacking they could sail closer to the wind, and by positioning himself upwind Drake forced enemies to fight at a disadvantage.

Drake could therefore outmaneuver his opponents, choose the time and place of engagement, and avoid being brought to close-quarters where Spanish boarding tactics were most effective. At Gravelines (1588), Drake's squadron, like others in the English fleet, exploited superior speed and maneuverability to harass the Spanish Armada's formation with long-range broadsides, crippling several ships and contributing to the fleet's fatal disarray.

Drake's early raids in the Caribbean and on the Spanish Main also honed his skills in deception and surprise. He frequently employed feints to draw enemy forces out of position, followed by flank attacks or ambushes.

In his 1586 assault on Santo Domingo, Drake landed troops at night away from the main defenses, then coordinated a two-pronged attack with a combination of land and sea forces. After landing troops at night and bombarding the city, he demanded ransom and, when refused, burned parts of the settlement until his terms were met. This operation exposed the vulnerability of Spanish colonies and forced Spain to reconsider its defensive posture in the New World.

Line-ahead Broadside Tactics

Rejecting the Mediterranean traditional approach of grappling and boarding, Drake pioneered using the line-ahead formation for stand-off gunnery firing from a distance of 200-600 yards. This inflicted damage on enemies without risking his own ships to their boarding parties. This gunnery-at-a-distance fighting was a departure from the melee-style, chaotic boarding tactics of the time, and it became the standard Royal Navy fighting method for the next 250 years.

For the line-ahead formation, he organized his ships into single-file lines and delivered sustained broadsides that devastated Spanish galleons. English ships, under Drake's guidance, were equipped with heavier gun-batteries with longer-range than Spanish galleons or galleasses, and his crews were drilled to deliver rapid, accurate broadsides. In the first pass his ships shot to cut enemy rigging and clear decks, while on the second and third passes he aimed to sink or cripple.

This tactic proved devastating at Gravelines, where Drake's squadron formation and its cannon fire shattered Spanish rigging and morale, and with aid from Lord Howard's fire-ships he contributed to the Armada's defeat.

Fireship Attacks

Drake's innovative use of fireships – vessels set ablaze and sent into the midst of the anchored Spanish fleet to scatter them – was a signature tactic that sowed chaos and confusion among the Spanish ranks, forcing them to break formation and scatter. It commonly triggered retreats that created opportunities for attack.

For this tactic, old vessels loaded with combustibles were sent drifting into Spanish formations. The burning ships forced the Armada to cut anchor and scatter under panic, thereby leaving gaps for English ships to exploit. This was revolutionary in the Age of Sail, as it introduced a low-cost, high-impact tactic that influenced later admirals like Nelson.

The most prominent example of the fireship strategy is when Drake, with a small squadron, sailed into the heart of Spanish naval power at Cadiz harbor, burned 37 Spanish ships, and captured or burned vast stores of supplies before withdrawing. This attack, previously mentioned, delayed the Spanish Armada by a year, buying England critical time to prepare. The raid forced Spain to divert resources, and constituted a strategic coup that "singed the King of Spain's beard" while demonstrating the value of striking first.

At Gravelines, eight fireships were deployed, scattering the Armada's tight formation, and enabling English ships to attack the disorganized galleons, sinking 5. Drake's 20 ships also attacked Vigo (1589) using rapid maneuvers to also burn supplies and evade pursuit, weakening Spanish coastal defenses. During the Lisbon Raid (1589) as part of the English Counter-Armada, Drake's fireships also disrupted Spanish preparations, burning supply ships and delaying reinforcements.

Intelligence and Reconnaissance

Drake's reliance on intelligence and reconnaissance, often overlooked, was a critical tactic that informed his raids and battles. He gathered information from captured sailors, local informants, and his own scouting, which together helped to ensure precise targeting of vulnerable targets, avoid traps, and exploit Spanish disorganization. One might say that his victories began with his spies.

This approach was revolutionary in an era when intelligence was rudimentary, and gave Drake a strategic edge over Spanish commanders. Essentially, his reconnaissance helped to counter the advantages of Spain's larger fleets, ensuring tactical surprise.

For instance, Drake's scouts identified Cadiz as a poorly defended supply hub, which enabled his surprise raid in 1587 that destroyed 24–30 ships. Captured Spanish sailors revealed treasure convoy schedules, allowing Drake to ambush Panama's port of Nombre de Dios. And of course, Drake's scouting ships tracked the Spanish Armada movements, enabling rapid strikes at Gravelines.

Revolutionary Impact and Lasting Legacy

Drake's tactics were revolutionary because they blended piracy-inspired privateering, intelligence gathering, rapid sail-powered maneuvers, and amphibious warfare, which challenged Spain's reliance on large, slow fleets and traditional battles. His tactics are studied at naval academies today because they emphasize adaptability over rigid doctrine (a lesson we see time and again in the lives of great naval commanders), the importance of disrupting enemy logistics, and the benefit of psychological intimidation before engagement.

His fireship tactics remained a staple of naval warfare for centuries. His commerce raiding anticipated modern economic warfare, influencing admirals like Hayreddin Barbarossa and Robert Blake. His daring raids proved that a small, agile force could harm an entire empire by operating on a motto sometimes attributed to Drake: "The sea rewards the swift and cunning." His integration of intelligence and amphibious operations also set a precedent for modern hybrid warfare.

Drake's win-loss record (~4 major wins, 1 minor loss at San Juan, Puerto Rico, 1595) reflects his ability to defeat larger Spanish forces through speed, deception, and surprise. His victories at Cadiz, Cartagena, and against the Spanish Armada weakened Spain's naval dominance, paving the way for England's maritime rise.

Drake's career, marked by a string of victories against the Spanish Empire, not only elevated England to the status of a global maritime power but also revolutionized naval tactics and strategy. His vision anticipated the later doctrines of sea power articulated by theorists like Alfred Thayer Mahan, and his campaigns laid the groundwork for the emergence of a professional Royal Navy.

CHAPTER 14: ANDREA DORIA (1466–1560)

Admiral Andrea Doria (1466–1560), the preeminent Genoese statesman and naval commander of the 16th century, was a master tactician whose strategies dominated the Mediterranean during the turbulent 16th century. He secured a string of victories over Ottoman Turks, Barbary pirates, and rival European powers that preserved Genoa's independence and bolstered the Holy Roman Empire's Mediterranean dominance. Doria's tactics were instrumental in achieving Christian maritime supremacy, Ottoman containment, and Genoese commercial prosperity.

Doria's genius lay not only in his battlefield prowess but in multiple areas: his mastery of coalition diplomacy, his use of disciplined galley fleets and innovative artillery, and his ability to adapt land-warfare principles to amphibious operations. Like other great admirals, Doria understood that naval power served broader political objectives, and he used maritime dominance to secure Genoese independence and Mediterranean trade routes.

Doria's career was forged in the fiery competition of the 16th-century Mediterranean, where empires and city-states clashed over every mile of sea lane. During his day, the Ottoman Empire, Spain, France, and the Italian city-states all vied for control of vital sea lanes. Genoa, Doria's homeland, was a small but wealthy republic whose survival depended on maritime trade and security so Doria accordingly recognized that Genoa's fortunes would be determined at sea. He committed himself to building a fleet, not just for defense, but for offensive action – raiding, blockading, and projecting Genoese power far beyond its shores.

Admiral Doria's strategic vision was remarkable for its flexibility. He was a master of shifting alliances, serving the French crown before switching his allegiance to Charles V of Spain when it better served Genoa's interests. This political adaptability was matched by a willingness to employ new tactics and technologies in pursuit of Genoese security.

As a *condottiero* (professional military leader who commanded mercenary armies) turned admiral, Doria had many of the faults of his profession: he was considered greedy, conceited, vindictive, unscrupulous, cruel, and authoritarian. Yet he was also a fearless and untiring military commander who was endowed with outstanding tactical and strategic talents.

Doria's signature tactics – commerce raiding and anti-piracy operations, defensive fleet withdrawals, amphibious assaults, and his political acumen together with leadership through delegation – set him apart from predecessors like Don Álvaro de Bazán, who prioritized galley battles, and peers like Hayreddin Barbarossa, his Ottoman rival. Unlike many of his contemporaries, Doria did not rely solely on brute force or numerical superiority. Instead, he perfected a blend of tactical innovation with disciplined seamanship that allowed him to outmaneuver larger Ottoman fleets and rival European powers.

Adaptive Fleet Composition

Doria's fleets favored oar-powered galleys, which, in calm winds, outmaneuvered larger sailing ships. However, Doria optimized his fleet's composition for specific missions by combining galleys, galleons, and support vessels in flexible formations. He was among the first to effectively integrate newer sailing vessels with traditional galleys, maximizing the advantages of both types of ships in his fleets.

While many Renaissance fleets suffered from starvation and disease, Doria ensured that his ships were constantly resupplied, which extended their operational range since he could then keep fleets at sea longer than his rivals. He secured key bases like Genoa, Naples, and Messina to allow rapid fleet repairs and reinforcement.

Line-Abreast Artillery Formations

Unlike predecessors who relied primarily on ramming and boarding tactics, Doria also organized his galleys in disciplined line-abreast formations that created overlapping fields of fire. This maximized his artillery effectiveness while allowing for each galley to protect its neighbor's flanks. The formation essentially maximized the destructive power of bow-mounted cannons while preventing enemy galleys from attacking the vulnerable sides of ships.

The tactical imperatives of the line-abreast formation were more compelling for the more heavily manned and armed Western galleys because in a melee the quicker and more maneuverable Muslim vessels could overwhelm them individually from the sides. As a result, the line-abreast formation, in which each galley protected its neighbor's flanks, became tactically dominant. In fact, line-abreast formations became standard Mediterranean galley tactics, influencing naval warfare for decades. If appropriate intervals were maintained, full-sized galleys could not slip through the formation, and smaller vessels that could generally do so lacked sufficient height and numbers of men for successful boarding operations.

While most Mediterranean admirals relied on massed galley charges, Doria also perfected targeting strikes against enemy flagships, knowing that decapitating command structures led to disarray. An example of this strategy was seen at the Battle of Pianosa (1519), where Doria's fleet broke through the center of the Barbary corsair line to isolate and capture their admiral's ship. He put into effect an observation of Alfred Thayer Mahan: "The quickest way to win a sea battle is to make the enemy's fleet leaderless."

Multiple Artillery Barrages and Bombardment Innovations

Armament in Andrea's time consisted of a bow cannon, aimed by pointing the vessel herself, and several smaller cannon along the sides of the galley. The accepted tactic of engagement was to hold fire of the bow cannon, which fired a 35- to 50-pound stone or lead ball, until just before crashing with the galley's bow boom against an enemy vessel.

Doria extended this to multiple salvos that devastated enemy formations before any boarding action. Doria, like many commanders of his era, employed ruses such as flying false colors when approaching unsuspecting foes to gain an advantage. Then he would implement multiple-shot sequences instead of the traditional single broadside before boarding.

In the summer of 1532, Doria conducted one of the most skillful naval operations of the 16th century by leading a Spanish-Genoese fleet of 48 galleys and 35 support vessels against Ottoman positions in the Aegean, aiming to divert Ottoman forces from Central Europe by relieving pressure on the Habsburg forces in Hungary. Departing from Messina, Doria assaulted coastal forts like Coron by combining naval gunfire with amphibious landings – using shipboard artillery to suppress defenses while troops scaled walls with rigging-lowered ladders.

This marked an early example of coordinated naval bombardment in siege warfare, and one of the earliest documented uses of mast-mounted siege cannons – a precursor to later innovations in naval bombardment.

Another example of superior artillery tactics was at the Battle of Antipaxos (1537), where Doria's fleet – equipped with improved bow artillery and flexible gun placements – reportedly defeated an Ottoman force, capturing several galleys with minimal losses.

By leveraging disciplined gunnery to disrupt enemy formations before closing, Doria demonstrated the tactical edge of integrated naval firepower in galley combat, although boarding actions still remained critical. This engagement foreshadowed the Mediterranean's shift toward gun-centric warfare, even as true broadside tactics awaited later sailing ships.

Commerce Raiding and Anti-Piracy Operations

Doria's commerce raiding and anti-piracy operations targeted Ottoman and Barbary pirate shipping, aiming at weakening their economies and protecting Genoese trade. Unlike traditional admirals who sought fleet battles, Doria's raids were defensive measures because they aimed at disrupting enemy logistics, mirroring Barbarossa's strategy of targeting an enemy's seaborne trade and supply lines rather than engaging in direct battles.

At the Battle of La Goulette (1512), Doria conducted an anti-piracy operation where his 12 galleys sacked Barbarossa's base, destroying ships and supplies. This raid weakened the pirate trade and Ottoman piracy.

At Pianosa (1519), Doria's 8 galleys defeated a Turkish raiding fleet, capturing ships and disrupting Ottoman trade routes and commerce. This move was yet another decisive blow to Turkish piracy that secured Genoese shipping lanes. Also, at the Battle of Girolata (1540), Doria's nephew captured Dragut, a key pirate, which disrupted Barbary raids. However, this victory was only temporary because Dragut was later ransomed and went on to become an even bigger threat.

Yet another example of commerce raiding occurred in 1524, when Doria's fleet raided French supply lines, forcing the imperial siege of Marseille to lift and protecting Genoese trade.

Doria's anti-piracy sweeps secured the safety of Genoese commerce, ensuring Genoa's economic stability by making the Mediterranean safe for its merchants. Raiding an enemy's shipping and commerce drained the opponent's resources, which forced Ottomans and pirates to divert their ships to protect convoys.

Defensive Fleet Withdrawals

Doria's practice of defensive fleet withdrawals, retreating to avoid decisive losses against superior forces, was a hallmark of his strategic restraint. Unlike peers like Barbarossa, who favored aggressive engagements, Doria viewed retreat as a tool to maintain long-term superiority because it preserved his fleet for future battles, a pragmatic approach in the resource-scarce Mediterranean. By withdrawing from unfavorable battles, Doria avoided catastrophic losses against superior forces, thus maintaining Genoa's naval strength against the larger Ottoman and French fleets. His retreats also sometimes lured enemies into overextending, creating opportunities for counterstrikes.

As examples, Doria's 12 galleys retreated from Barbarossa's base at La Goulette (1512) after sacking it, avoiding a larger Ottoman response. This was a calculated necessity that preserved his fleet for future raids.

At the Siege of Naples (June 1528), Doria's fleet actually switched sides during the siege and withdrew from French naval forces, again preserving his galleys. He switched his allegiance to Emperor Charles V because of disillusionment with French leadership. With support from the Imperial fleet, Doria then drove out the French garrison from Genoa and liberated it from foreign control.

Doria's defection and military action ended French dominance in Genoa. He then restructured Genoa's government, establishing a more stable republic under his guidance. He became the de facto ruler of the city, though he refused to take formal power. His alliance with Charles V consequently ensured Genoa's autonomy and prosperity for decades.

At Preveza, Doria faced Barbarossa's 122 galleys with 195 Holy League ships, but withdrew after initial skirmishes. Doria's caution and failure to press an advantage led to a decisive Ottoman victory, cementing their naval dominance for decades. However, his preservation of the Holy League fleet prevented its total annihilation.

Amphibious Assaults

Doria's mastery of amphibious assaults, combining coordinated naval bombardment with infantry insertion to capture key strongholds, was a cornerstone of his success. Unlike predecessors who focused on sea battles alone, Doria integrated galleys with troop landings to seize strategic ports, a tactic rare in the 16th century.

This seamless fusion of sea and land assault was groundbreaking, a primitive blueprint for the combined-arms operations of a much later age. Amphibious assaults allowed Doria to strike at the heart of enemy power – ports and fortresses – while minimizing risks. The amphibious assaults exploited the mobility of galleys to land troops swiftly, bypassing fortified defenses and catching enemies off-guard. Doria's precise timing and coordination ensured rapid victories with minimal losses, securing strategic bases with fewer casualties.

Doria's amphibious campaigns were central to restoring Genoa as an independent Mediterranean power. Several examples of his amphibious assaults include the liberation of Genoa (1528) that restored its independence, Coron Capture (1532) and the conquest of Tunis (1535).

Earlier I explained that Doria switched sides during the Siege of Naples, and then later began an amphibious assault on Genoa in an attempt to drive out the occupying French garrison, which was weakened at the time by an attack of the plague. Doria led 12 galleys and 1,000 troops to expel the French forces from the city, defeating the local French naval support and landing at night to overwhelm the weakened garrison. Doria coordinated naval artillery suppression with synchronized multi-point infantry landings for the troops, overwhelming defenders who couldn't respond to simultaneous threats from sea and multiple landing sites.

At Coron, Doria's 27 galleys landed troops at the Ottoman-held city to capture the port in a swift assault. His ships used mast-mounted cannons to bombard walls and ship-mounted ladders for scaling. This surprise attack, using technology adapted to the needs of the moment, captured Coron's fortress and the city, disrupting the Ottoman supply lines in Greece.

At Tunis, sharing command with Álvaro de Bazán, Doria's fleet of 74 galleys landed 20,000 troops against Barbarossa's defenses, capturing Tunis and 42 ships. His amphibious coordination turned a naval expedition into a land victory, weakening Ottoman North Africa.

Command Structure & Coalition Fleets

Doria's organizational structure delegated tactical authority to subordinates while maintaining strategic control of engagements. This decentralized tactical command, with Doria providing centralized strategic direction, has been seen elsewhere and became a model for later naval organizations. By delegating full command authority to subordinates during missions, he empowered captains for independent actions which allowed rapid responses in fluid battles.

Delegating full command authority to subordinates was revolutionary because it fostered initiative in an era of rigid hierarchies, and it allowed Doria to conduct multiple simultaneous operations while maintaining tactical flexibility. Doria's leadership, delegating to trusted subordinates like Filippino Doria, ensured tactical flexibility as well as loyalty. Filippino's success was but one example of Andrea's ability to choose able subordinate commanders and indoctrinate them quickly into his methods.

Doria also forged alliances that lasted through the Italian Wars, and circumvented political rivalries among coalition partners to maintain a unity across Italian, Papal, and Habsburg factions. He commanded diverse forces under Habsburg, Genoese, Venetian, Papal, and Maltese flags.

Commanding a coalition fleet, Doria at one time blockaded Genoa against Spanish forces. In a fierce four-hour battle, he led the right flank, personally engaging enemy galleys and directing the action. His ability to coordinate a multinational fleet and delegate authority to trusted subordinates was one of the keys to victory.

Doria's Revolutionary Impact and Lasting Legacy

Doria's win-loss record (~4 major wins, 1 loss at Preveza) reflects his ability to defeat larger Ottoman and French fleets through cunning and restraint. His victories at Pianosa, Tunis, Coron, and Genoa secured Mediterranean trade routes and Genoa's independence. His influence persisted in Genoese and Spanish naval strategies, shaping the Mediterranean balance of power into the future.

Andrea Doria did not win every battle, but his tactical innovations protected vital Mediterranean trade routes. His life's work was a single, grand strategy with three goals: asserting Christian control of the sea, checking Ottoman expansion westward, and bolstering Genoese commercial prosperity.

Doria successfully operated in the transition period between two eras of naval combat, employing the traditional boarding tactics of the past while pioneering the gun-centric warfare of the future. His tactics were revolutionary because they integrated line-abreast and artillery innovations, amphibious warfare, strategic restraint, and a decentralized command structure into a cohesive strategy, challenging the galley-era focus on direct fleet battles. His amphibious assaults foreshadowed modern joint operations while his defensive withdrawals preserved Genoa's navy.

Doria's commerce raiding anticipated Drake's privateering strategy and his psychological warfare set a standard for intimidation, while his command delegation set a model for later admirals like Nelson. His successes in protecting Genoa proved that strategy, which requires thinking, triumphs over brute force. His successes in protecting Genoa proved that strategic acumen and political cunning could protect a city-state where brute force alone would have doomed it.

CHAPTER 15:
THOMAS COCHRANE (1775–1860)

Admiral Thomas Cochrane, 10th Earl of Dundonald (1775–1860), was one of Britain's greatest sea captains, and the inspiration for naval fictions featuring fearless commanders against odds. He was an audacious naval officer who achieved spectacular victories against heavy odds through an innovative use of deception and aggressive tactics that consistently overwhelmed opponents. He earned the moniker "*Le loup des mers*" ("Sea Wolf") from Napoleon, for this single image captured his aggressive speed, use of deception and tendency to scare opponents.

His service spanned the Royal Navy and the navies of Chile, Peru, Brazil, and Greece, and was marked by extraordinary triumphs that liberated nations and humbled empires. His track record of being undefeated in major actions includes key winning engagements in the Napoleonic Wars and Latin American independence movements.

Cochrane's approach to naval warfare was characterized by boldness, deception, and the exploitation of maneuverability. Unlike many contemporaries who relied heavily on brute force and line-of-battle tactics, Cochrane emphasized speed and surprise along with psychological warfare. He understood that smaller, faster ships could outmaneuver larger opponents if handled with skill and audacity. His tactics often involved close-quarters engagements, feints, and boarding actions, where he turned the enemy's numerical or firepower advantage into a liability. Modern scholars like David Cordingly, Andrew Lambert, and Robert Harvey agree that Cochrane's principles reshaped naval doctrine.

Cochrane's special tactics – commerce raiding, strategic deception, fireship attacks – set him apart. Where Nelson sought decisive fleet engagements, Cochrane mastered asymmetric warfare. Cochrane's methods were groundbreaking for their disproportionate impact from his limited resources, transforming naval combat in the Age of Sail.

Commerce Raiding and Privateering

Unlike traditional admirals focused on fleet battles, Cochrane, a licensed privateer, prioritized disrupting trade to weaken an enemy's naval capabilities. Cochrane's mastery of commerce raiding and privateering – targeting merchant ships and convoys under letters of marque – crippled foes economically by taking their gold while enriching his cause. This was a shift in the very objective of war at sea, which now moved from territorial clashes to economic attrition designed to bankrupt an enemy's will and means to fight. It elevated privateering from piracy to strategic warfare. Cochrane influenced independence movements and modern economic blockades by demonstrating how small forces could undermine empires.

Raiding commerce drained enemy resources, sowing economic chaos and forcing the diversion of warships to protect convoys. Cochrane's small, fast ships outran larger fleets, achieving an outsized impact with minimal risk. His operations not only funded rebellions but also demoralized enemies, with his captured prizes boosting his crews' morale and finances.

During the Mediterranean Raids of 1800–1801, which exemplified economic warfare at its finest, Cochrane commanded *HMS Speedy* and captured over 50 Spanish merchant vessels, disrupting trade and forcing Spain to commit warships to convoys. His raids yielded prizes worth £75,000, weakening Spanish logistics and demonstrating how privateering could sustain prolonged campaigns.

In the fight for Chilean Independence (1818–1822), Cochrane was commander of the Chilean Navy and raided Spanish convoys along the Pacific coast, capturing ships and supplies that funded the rebellion. His operations isolated Spanish garrisons, crippling their ability to resupply, fight, and felling Spain's South American empire.

During the Peruvian Liberation (1820), Cochrane's raids on Spanish supply lines supported San Martín's army, capturing vessels that sustained Peruvian independence.

During the Brazilian War of Independence (1823–1825), Cochrane's squadron raided Portuguese convoys in Bahia, capturing merchant ships and forcing evacuation. This disrupted Portugal's transatlantic trade, hastening Brazil's freedom.

These few examples show how commerce raiding worked to create a multiplier effect: each captured ship not only deprived the enemy but also armed and financed Cochrane's forces, allowing sustained operations against superior numbers while pulling away bricks from the walls of empire. As we have repeatedly seen, naval warfare is incredibly tied to the economics of naval operations and the budgets of host nations.

Aggressive Frigate Tactics and the Speed Emphasis

Cochrane's tactics consistently emphasized speed and maneuverability as force multipliers. He always aimed to leverage a smaller ship's agility with the element of surprise rather than relying on superior firepower. He operated on a fundamental principle: speed delivered surprise, and surprise delivered the opportunity for a precise, crippling blow. Applying this principle enabled him to dictate the terms of engagement, choosing when and where to strike.

Cochrane avoided prolonged artillery duels that favored larger ships and instead sought to close rapidly to board or disable enemy rigging, rendering the ships immobile and vulnerable. This approach was a marked departure from the rigid line-of-battle tactics dominant in the 18th and early 19th centuries, which emphasized broadside exchanges in linear formations.

While most captains engaged in long-range duels, Cochrane preferred to sail directly into the enemy's broadside arc, hold fire until the last moment, then unleash a devastating close-range volley. In the naval warfare of the time, the first well-placed broadside often decided the battle, and Cochrane used the subterfuge of disguise to accomplish this, approaching under false colors before revealing his intent.

This type of sudden, unexpected and overwhelming attack was used in the battle between *HMS Pallas* and the French 74-gun *Minerve* (1806), where Cochrane's aggressive charge closed the distance rapidly, neutralizing *Minerve's* superior guns through speed and surprise. He captured the ship after a fierce exchange, showcasing how frigate agility could overcome a ship-of-the-line.

His flexible, aggressive style proved that smaller vessels could punch above their weight through superior handling and surprise rather than relying on superior firepower, which foreshadowed modern naval doctrines valuing mobility and initiative. Cochrane believed and proved that psychological manipulation and audacity could unbalance even vastly superior forces.

Close-Quarters Boarding

The principle of naval fighting during his day was that when guns fail, the blade decides – close the gap, and own the deck. Cochrane's tactic of close-quarters boarding, closing at close range to grapple and board enemy ships, exploited his crews' discipline and speed to overwhelm larger forces. Unlike distant gunnery, Cochrane favored hand-to-hand combat to minimize damage to prizes, maximizing his economic gains by capturing other vessels intact. He would regularly concentrate 2-3 times the normal boarding party on one small part of the enemy deck. His meticulous planning and training of his crew ensured that every movement was executed with precision, minimizing casualties and maximizing effect. He remains the only captain in history who could reliably take ships five or six times his own force by boarding alone.

Cochrane's capture of the Spanish xebec *El Gamo* (1801) while commanding the tiny *HMS Speedy* remains legendary. *HMS Speedy* was just a sloop with only 14 four-pounders and 54 men, and was going against a ship three times its size with 32 guns and a crew of 319 men. Cochrane employed clever subterfuge by sailing under false colors – first American, then British once close – to confuse the enemy, and his emphasis on speed allowed his tiny sloop to sail close enough that the *El Gamo* could not depress its guns to fire on *Speedy's* hull. His gunners disabled rigging first, immobilizing the frigate and forcing the Spanish into a desperate boarding attempt, which Cochrane then turned against them by pulling away and firing on the boarding parties before ultimately ordering his own men to board and capture the larger vessel.

Cochrane's men repeated a speedy capture on the *Esmeralda* (1820) where they boarded the Spanish frigate in Callao harbor, capturing it in 15 minutes. In 1801, *HMS Speedy* also boarded and captured multiple Spanish frigates in rapid succession.

Deception, Bluffing and Stealth

Unlike peers, Cochrane commonly used psychological tricks to win battles. He mastered the art of strategic deception by using false flags, decoy lights, disguise and audacious ruses to mislead, evade, and outwit his enemies. These deceptions, inspired by corsairs, allowed him to approach enemy fleets without alarm and preserved his fragile ships while emboldening crews to engage under more advantageous circumstances.

Cochrane's deception and bluffing often outwitted enemies without firing a shot. On one occasion, he flew false flags and claimed plague aboard his ship, prompting the Spanish to break off a boarding attempt. During the daring *El Gamo* capture, by initially flying American colors (legal under contemporary *ruse-de-guerre*) until switching to the British flag, he avoided early enemy fire. In one evasion (escaping from the French Squadron in 1800), he even attached a candle to a floating barrel to mislead pursuers and escaped in darkness; the enemy pursued the light while *Speedy* slipped away.

In the Brazilian Maranhão Bluff (1823) case, Cochrane sent a single ship with false dispatches claiming a large fleet approached, bluffing the Portuguese garrison into surrendering. This bluff captured the province without battle. During the Belém Surrender (1823), Cochrane bluffed Portuguese forces with fabricated reports of Brazilian reinforcements, also securing surrender without combat.

As to stealth, Cochrane's strategic ingenuity extended beyond single-ship actions to fleet operations. In the Chilean War of Independence, he masterminded the nighttime boarding and capture of the Spanish frigate *Esmeralda* in 1820. Understanding that a direct confrontation would be costly, Cochrane devised a plan centered on stealth and surprise. He handpicked a strike force of 240 sailors and marines, drilling them relentlessly for the operation. To lull the enemy into complacency, he staged a feint by having two of his ships visibly weigh anchor and sail away, while he anchored his flagship out of sight behind an island. Under cover of darkness, his forces then boarded and captured the *Esmeralda* with minimal losses. This operation showcased Cochrane's mastery of deception, timing, and rigorous preparation, which allowed him to neutralize a superior enemy force without a protracted battle.

In the Greek War of Independence, he employed deception by sometimes disguising his flagship as a commercial vessel to approach enemy positions. For example, the frigate *Hellas* flew the Sardinian flag and acted as a merchant escort to mask its true purpose.

This tactic defied convention in an era when naval engagements often depended on clear identification of friend or foe. As one would expect, Cochrane's use of disguise allowed his fleet to approach enemy ports and positions with reduced risk of detection, enhancing the element of surprise. Although an operation at Alexandria in 1827 lost the element of surprise due to an Egyptian patrol's intervention, the concept itself was forward-thinking and influenced later naval deception practices.

Fireship Attacks and Explosion vessels

Admiral Cochrane's innovative fireship attacks, setting ablaze vessels or using them as explosive vehicles to disrupt enemy formations, was another tactic that created panic and openings for assault. Unlike the standard broadside exchanges of his time, Cochrane's fireships were low-cost tools that forced larger fleets into disarray, exposing them to follow-up assaults, and often decided battles before conventional combat could even begin. Cochrane's precise wind and tide calculations maximized the intended chaos. This fireship tactic profoundly shaped the strategies of naval blockades and assaults.

Cochrane's most famous use of fireships was at the Battle of Basque Roads (1809), which was a dramatic naval engagement during the Napoleonic Wars. Cochrane's sent fireships and explosion vessels (packed with 1,500 barrels of gunpowder) into the heart of the French fleet anchored near Aix Roads on the Biscay coast of France, which sunk 4 ships while also forcing the French to cut their cables and run 11 ships aground in panic. This disrupted Napoleon's invasion plans and showcased the effectiveness of unconventional naval tactics. During the Valdivia Capture (1820), Cochrane used fireships to ignite Spanish forts, causing panic and enabling his 300 men to capture Valdivia, Chile.

Coastal Raids and Amphibious Operations

Cochrane's coastal raids and amphibious operations targeted ports and forts, blending naval strikes with land assaults as one would expect. Unlike fleet-focused admirals, his raids disrupted logistics while his amphibious landings captured key positions, and his speed ensured escape.

For instance, he targeted supply lines during the Chilean and Peruvian Campaigns (1818–1825) where his raids crippled Spanish logistics. In 1820, Cochrane's 300 men raided seven forts, capturing Valdivia, Chile with surprise landings. In the Greek War of Independence (1827), he experimented with steam-powered vessels (a then-novel technology) for coastal bombardment. In 1823, his coastal raids forced the Portuguese evacuation of Bahia, and in 1820 an amphibious boarding captured the *Esmeralda* at Callao Harbor.

A Revolutionary Impact and Lasting Legacy

Cochrane's legacy endures in naval history, fiction, and doctrine. His win-loss record (~5 major wins, no losses in pitched battles, but numerous other single-ship victories) reflects his ability to defeat larger forces through ingenuity, and his small forces regularly achieved a disproportionate impact. His tactics were pioneering because they combined deception, psychological warfare, and unconventional tools like fireships and floating bomb explosion vessels. His privateering and bluffs influenced independence wars, while his boarding and raids inspired fiction by O'Brian and Forester. Cochrane defied naval orthodoxy, regularly challenging line-of-battle norms, and personified aggressive innovation.

Naval historians and strategists recognize Cochrane as a forerunner of modern naval warfare because he blended scientific tactics with psychological strategies. Though politically controversial – he was dismissed from the Royal Navy due to a stock fraud scandal but later reinstated – his tactical genius redefined sea warfare. His emphasis on speed, stealth, and rigorous preparation allowed him to win battles that seemed impossible on paper. He was a naval commander who won battles not by brute force, but by outthinking, outmaneuvering, and out-daring his enemies.

CHAPTER 16: FYODOR USHAKOV (1744–1817)

Admiral Fyodor Ushakov (1744–1817) commanded Russia's Black Sea and Mediterranean fleets with flawless success during the Age of Sail. In the course of 43 naval engagements under his command, he never lost a single ship or a battle, which is a distinction rare in naval history.

This extraordinary, unblemished record distinguishes him even among the pantheon of great naval commanders. Ushakov's consistent success, often against numerically superior forces, certainly wins him a place among the best naval commanders of all time and demands a thorough examination of the strategies and tactics that underpinned his winning track record.

Ushakov was a naval commander who defied the conventional wisdom of rigid 18th century line-of-battle doctrine, preferring rapid, decisive action over rigid adherence to tradition. He won battles by developing and refining tactics and strategies that produced unprecedented success. His innovations earned him honors from both Russian and Ottoman empires, and even sainthood in the Russian Orthodox Church.

His revolutionary strategies – broken line formations and flanking, reserve squadrons for decisive strikes, close-quarters maneuvers and artillery integration, amphibious operations, psychological warfare through morale, and leadership infused with faith – set him apart from predecessors like Maarten Tromp, who refined line tactics, and peers like Horatio Nelson, whose aggressive actions contrasted with Ushakov's calculated precision.

Ushakov created a new paradigm of naval warfare that combined aggressive maneuvers with innovative formations, disciplined gunnery, and rapid decision-making. His victories against the Ottoman Empire in the Black Sea and the Mediterranean not only expanded Russian naval power but also established doctrines that would influence future generations of naval strategists.

During Ushakov's lifetime in the late 18th century, naval warfare was still dominated by the line-of-battle tactic. In this formation, ships of the line would array themselves one after the other at regular intervals of about 100 or more yards forming a single column, and these enemies arranged in parallel lines opposite each other would engage by exchanging parallel bombardments at relatively fixed ranges in hours-long unhurried battle.

Adhering to a line reduced the risk of friendly fire, allowed concentrated firepower and enabled an admiral to maintain cohesion and control of his fleet. However, the rigidity of fighting in a line constrained initiative and often led to stalemates between evenly matched forces.

Ushakov rejected this doctrine outright. Instead of adhering slavishly to the linear template, he championed maneuver warfare on the sea. He combined fleet formations with continuous movement, breaking enemy lines and striking at the enemy flagship. This shifted the nature of naval engagements from passive fire duels into fast-paced combined arms operations at sea.

As Russian naval historian R. H. Cross summarized, "Ushakov's tactics did not discard the line, but subordinated it to maneuver – creating superiority in sectors of attack, rapidly deploying forces, and seizing tactical surprise." This fusion of order and opportunism was a radical departure from the sclerotic norms of his era.

Aggressive Maneuver Warfare

As a geopolitical background, the late 18th century was a period of profound geopolitical transformation, particularly for the burgeoning Russian Empire under Catherine the Great. Russia was rapidly expanding its influence, with the Black Sea and Mediterranean becoming crucial theaters for projecting power and securing vital ice-free ports. Naval power was indispensable to these ambitions.

Within this dynamic environment, naval warfare was largely dominated by the rigid "line of battle" that was considered the ideal combat formation as articulated by theorists such as the Portuguese Fernão de Oliveira as early as 1555. However, this method of fighting frequently led to indecisive engagements particularly when opposing fleets were of comparable strength. These were often indecisive battles of attrition because the objective became that of disabling enemy ships by targeting their masts and rigging rather than outright destruction, which allowed damaged vessels to eventually withdraw from the fight, get repaired, and one day fight again. It was within this context of established, yet often ineffective naval norms that Ushakov emerged as a true innovator.

Breaking an enemy's line could cause its collapse, but conventional doctrine made it difficult to mass forces for such a break, as it required violating the sacred order of the line itself. Ushakov rejected this passive formation in favor of dynamic, aggressive maneuvers by allowing his captains to maneuver freely within unified marching orders that allowed both maneuver and fire, and which gave them freedom to focus attacks where they mattered.

Ushakov's usage of unified marching and fighting orders, and his resolute closing to close quarters against enemy forces, meant his fleet could transition rapidly from a cruising formation directly into battle, minimizing the time and complexity required for traditional, time-consuming pre-battle deployments. By minimizing the time and complexity involved in forming a battle line, Ushakov gained a critical temporal advantage over his adversaries, often enabling him to achieve tactical surprise against an opponent expecting more conventional maneuvers. Essentially, the speed with which he could commit his forces prevented the enemy from fully forming their lines or reacting effectively to his aggressive approach.

Ushakov's maneuvers did not at all exclude line tactics as a battle formation, but the line was not the only way of fleet formation and it was completely subordinated to the maneuver. Unlike predecessors like de Ruyter, who perfected the line, what Ushakov did was draw from Suvorov's land tactics to create fluid "attack columns."

Ushakov's signature tactic became broken line formations – dividing his fleet into squadrons for independent maneuvers and allowing flanking attacks that encircled enemies. His broken lines confused enemies while preventing concentrated fire, enabling Ushakov's squadrons to target weak points like rear guards. His galleys' speed ensured flanks were turned before foes reacted.

Concentration of Force on Flagships

In the Age of Sail, fleet command and control was highly centralized and heavily reliant on flagships, which served as the primary communication hub for all the ships in the fleet. Disabling or destroying the flagship, especially the admiral's vessel, meant a direct "breach of command and control" for the enemy. Therefore, instead of distributing fire evenly across the entire enemy line, Ushakov concentrated force at decisive points, and most frequently at enemy flagships. This is a strategy we have seen repeated by many great admirals, and by great land generals too.

For Ushakov, a decapitation strike targeting enemy leadership was a high-risk, high-reward strategy far ahead of its time. By eliminating command ships, Ushakov aimed to dismantle the enemy's ability to fight, and create chaos that would make a coordinated response impossible. The loss of a flagship would not only shatter the enemy's confidence but often triggered a "general exodus" causing its entire fleet to rout. Therefore Ushakov decided to target the command ship whenever possible.

At the Battle of Fidonisi (1788), his first major engagement, Ushakov was in command of the Russian vanguard. He realized that the Turks by far outnumbered his troops so he seized the initiative by breaking from the traditional line, and used his vanguard to isolate and overwhelm the Ottoman flagship of the first column. At the Battle of Kaliakria (1791), Ushakov's 16 ships broke formation to flank 18 Ottoman vessels, sinking three and routing their fleet despite rough seas. At the Cape Tendra (1790), Ushakov divided his 10 ships into columns to flank 14 Ottoman galleys, capturing the flagship and scattering the rest.

Unlike the rigid parallel lines of his adversaries, Ushakov's fleet could pivot, envelop, or divide enemy forces. At the Battle of Kerch (1790), he used the mobility of his flanking squadrons of 10 ships to outmaneuver 14 Ottomans, forcing their retreat without losses. At the Corfu Siege (1799), Ushakov flanked French batteries with amphibious support, capturing the island.

Ushakov's practice was to mass his forces where it mattered most – an early example of the principle of force concentration. This tactic broke the enemy's coherence and achieved local superiority repeatedly.

In battle after battle, Ushakov's tactics allowed him to turn inferior numerical strength into decisive victories. His combination of precise gunnery and aggressive maneuvering (involving firepower and mobility) created a dynamic fighting style that traditional linear tactics could not counter because Ushakov could concentrate superior firepower at the point of his choosing while denying the enemy the ability to bring their full weight to bear. His tactics were an innovation in the theory of sea warfare.

Resolute Closing Integrating Gunnery Fire with Maneuver

While many admirals of his time favored long-range cannonades, Ushakov trained his crews for close-range, high-volume fire and boarding, thus ensuring flexibility in any combat scenario. His emphasis on gunnery drills ensured that his ships could deliver devastating broadsides at point-blank range to shatter the hulls of enemy ships.

The Russian tendency to aim at the Ottoman hulls would inflict damage to Ottoman gun crews. While the French and Ottomans preferred to target rigging, he focused on the hull aiming to put the "fighting machine, not the fighting man, out of action." This transformed Ushakov's fleet from a rigid, predictable formation into a flexible hammer.

While conventional naval wisdom also favored long-range engagements to minimize damage, Ushakov preferred to close rapidly to close quarters (often without reforming into battle line) to maximize the effectiveness of his gunfire and eliminate the enemy's ability to maneuver. At close range encounters, the Russian crew morale and prior training that integrated gunnery with melee gave them the edge. Ushakov effectively fused artillery with assault, thus revolutionizing sail combat in a way that bridged the galley and sail eras.

His fleet would aggressively approach the enemy, often to a distance less than 100 meters, which stood in sharp contrast to the French tendency to engage at longer ranges but with indecisive outcomes. By closing the distance rapidly, Ushakov denied the enemy time to prepare or capitalize on wind advantage, minimizing the enemy's ability to maneuver or disengage, and forcing a decisive engagement.

Ushakov's standard attack pattern could be summarized as follows:

1. Rapid approach under full sail to close distance to close quarters.
2. Concentrated fire on the enemy flagship and lead ships.
3. Combination of aimed artillery fire and maneuvering.
4. Maintain a reserve (Kaiser flag squadrons).
5. Relentless pursuit to prevent regrouping, chasing the enemy to its total destruction or capture.

As an example of this pattern in action, at the Battle of Tendra Ushakov closed within pistol-shot range before unleashing his broadsides, crippling the Ottoman fleet in under three hours.

Reserve Squadrons for Decisive Strikes

Traditional line-of-battle tactics often committed all available ships simultaneously to the single line, leaving little or no room for a tactical reserve. However, Ushakov created what was known as the "Kaiser flag squadron" – a mobile tactical reserve kept off the main fighting line that allowed him to commit fresh forces to the enemy's weakest point at the decisive moment. This was revolutionary – adapting the concept of land reserves to sea battles – and later became standard Russian naval doctrine.

The presence of a dedicated reserve provided Ushakov with a decisive advantage, enabling him to adapt to the unfolding battle and react to unforeseen developments (such as exploit breakthroughs, conduct strikes on exposed flanks or reinforce vulnerable points) or to deliver a crushing blow at a critical moment. This flexibility allowed him to overwhelm enemies who had already committed all their ships to the line of battle.

This adaptability, coupled with his aggressive closing tactics, ensured that once an advantage was gained it could be pressed to a decisive conclusion rather than devolving into an indecisive long-range duel. Ushakov's reserves exploited enemy fatigue, delivering knockout blows when foes committed fully while Ushakov's timing ensured minimal losses.

At the Battle of Kaliakria, for instance, Ushakov's reserves flanked the Ottoman rear and sunk three ships. At the Battle of Tendra, his reserves struck the Ottoman flagship and captured it. At the Battle of Fidonisi, his reserves repelled Ottoman attacks and forced their retreat.

Aggressive Pursuit: The "No Escape" Doctrine

Many Age of Sail naval battles often concluded with the defeated fleet retreating, frequently due to the inherent limitations of pursuit in sailing warfare. Ushakov's explicit emphasis on "aggressive pursuit" directly challenged this common outcome.

As with many land generals, he firmly believed in pursuing the enemy to its capture or total destruction so after battles his fleet would relentlessly pursue the fleeing enemy, inflicting further damage and ensuring their complete rout. He fought under one unbreakable rule: "A defeated enemy must not be allowed to escape. Every ship that can float must be captured or sunk. None must be allowed to escape and fight again tomorrow." Ushakov held back from any initial clash a permanent reserve squadron of his fastest, heaviest ships that was kept solely for pursuit.

The moment the enemy began to flee or surrender, Ushakov signalled "General Chase" and personally led the reserve at full sail. Individual captains were ordered to ignore formation and hunt independently, but forbidden to stop for prizes until the last enemy ship was taken or destroyed. Night, storms, or shallow water were no excuse; he pursued in all conditions (famously at Corfu and Tendra).

By destroying or capturing enemy ships, he permanently reduced an enemy's naval power, thereby securing control of key areas like the Black Sea and the Ionian Islands. This relentless pursuit also had a profound and lasting psychological effect on his opponents, contributing significantly to his reputation as "the terror of his country's enemies."

Superior Training with Rigorous Drilling

Alexander Suvorov, a legendary Russian land commander he admired, revolutionized Russian military training by emphasizing "speed and mobility, accuracy of fire, and the bayonet." Suvorov advocated for "realistic training for battle instead of the popular Prussian-style parade ground maneuvers," and stressed rapid, decisive attack while rejecting rigid linear fighting tactics. Suvorov had a famous motto, "Train hard, fight easy," which encapsulated his belief that "training made the injection of fury possible."

Ushakov was a staunch supporter of generalissimo Suvorov's principles of training for sailors and officers, and Suvorov's influence is seen in the parallel road that Ushakov put sailors upon. The core principles of speed, assessment, and hitting power from Suvorov's *Science of Victory* were clearly adapted by Ushakov to naval warfare. Admiral Zinovy Rozhestvensky attempted to copy Suvorov one century earlier but this ended in tragedy at Tushima, whereas Ushakov took the same Suvorov principles and made them work at sea. This intellectual adaptation of a highly successful land general's methods to naval fighting meant that Ushakov was applying a proven, aggressive military philosophy to naval warfare, which was largely stuck in a more conservative, attrition-based mindset.

Executing Ushakov's revolutionary tactics, particularly his aggressive maneuvers and preference for close-quarters combat, demanded exceptional skill, discipline and morale from his crews. Such complex and risky maneuvers could not be executed by poorly trained or demoralized sailors so rigorous training was required. Therefore, his emphasis on "morale and military skill of the personnel" was not merely a humane gesture but a strategic imperative.

To fill his needs, Ushakov's practical training of crews drilled both gunnery and boarding – ensuring flexibility in any combat scenario – and his personal care for their welfare fostered their loyalty and high morale. This strong foundation in training directly enabled the successful execution of his complex tactics, which allowed his fleet to perform skillful actions while withstanding the stresses of combat even against superior numbers.

Ushakov personally oversaw this rigorous training in seamanship, drawing inspiration from Suvorov's emphasis on practical readiness while emphasizing:

- Intensive gunnery practice
- Initiative-based leadership development
- Tactical flexibility over rigid adherence to regulations

For instance, Ushakov's fleets trained rigorously with unified signals and coordinated maneuvers. He personally innovated an extensive flag-signal system comprised of 130 signals that covered all battle situations. This sophisticated communication system allowed real-time coordinated battle movements and command adjustments unlike anything seen before –

capabilities that traditional navies lacked. His flag-signal system provided seamless communication and ensured that his fleet operated with a unified command of singular purpose and agility.

Additionally, Ushakov's command approach fostered a culture where subordinates were empowered to "act decisively within the commander's intent," where officers were given a clear mission and significant leeway in its execution. He was trusting their on-the-spot judgment. By empowering his officers to make on-the-spot decisions aligned with his overall intent, Ushakov effectively decentralized tactical execution. This allowed his fleet to react faster to enemy movements and exploit fleeting opportunities, thus overcoming the communication bottlenecks that plagued more formalist navies.

Amphibious Operations

Unlike many naval commanders who focused solely on sea battles, Ushakov excelled in joint operations, coordinating naval bombardments with the marine landings of troops under naval cover to seize islands. This was another concept ahead of its time, and Ushakov pioneered joint operations in the Black Sea.

His capture of Corfu – a heavily fortified island – where he landed 2,000 troops under bombardment demonstrated his ability to integrate fleet actions with ground assaults. At the Ionian Islands (1798–1799), his amphibious assaults seized multiple islands, and at the Siege of Anapa (1791) his naval support enabled the landing of troops and the capturing of the fortress.

In a coalition with Ottoman seamen, Ushakov blockaded, bombarded, and conducted amphibious assaults on Vido and fortified Corfu. His navy combined sea fire and troop landings to force the French surrender, which created the Republic of Seven Islands. These operations marked the birth of Russian naval amphibious doctrine.

Inspirational Leadership Under Fire & Faith-based Morale

Ushakov, like Nelson, led from the front, which instilled confidence in his men. He was famous for standing openly on the quarterdeck during enemy bombardment, never taking cover, and giving orders with deliberate, quiet confidence. This visible fearlessness reassured sailors that their commander possessed both courage and divine protection.

At the Battle of Cape Kaliakra (1791), he personally directed his flagship *Rozhdestvo Khristovo* into the thickest fighting, inspiring his fleet to a crushing victory. Before battles such as Fidonisi and Tendra, Ushakov gathered officers and sailors to pray together and seek divine protection. During the battle at Fidonisi, Ushakov broke some other rules of formal line tactics where the flagship was expected to be at the center of the line of ships.

Ushakov led by example, sharing the hardships of his crews. He strengthened morale with a deep sense of spiritual mission. After capturing enemy crews – especially Ottoman or French ships – Ushakov was known for humane treatment, refusing to execute prisoners or allow unnecessary cruelty. This compassionate conduct increased his moral authority among his own men, who saw him not only as a commander, but as a righteous leader guided by faith.

Ushakov also infused his fleet with the conviction that they were fighting not merely for Russia, but for righteousness, protection of Christian peoples, and defense of the innocent. His leadership, blending military acumen with Orthodox faith, boosted morale (reducing fear in battle) while inspiring unwavering loyalty.

Ushakov's Russian Legacy

Ushakov remains one of the few great admirals with an undefeated record, which underscores the practical effectiveness of his warfare tactics. By daring to defy stale conventions, Ushakov also achieved more victories in decisive fashion than most legendary naval commanders. He revolutionized how naval battles could be won and reshaped naval warfare for future generations. He blended line of battle with maneuver warfare, adapted formations to the battle, trained gunnery for accuracy rather than just volume, became a master of close-range decisive combat with that precision gunnery, pioneered amphibious cooperation by employing sea bombardment plus coordinated landings, would strike to sink an enemy flagship to cause chaos in the enemy command structure, maintained a mobile reserve as reinforcements or to exploit openings, and never allowed a wounded enemy to regroup. He also revamped naval training with a focus on artillery, seamanship, and boarding tactics.

Unlike contemporaries, Ushakov would also engage the enemy quickly without waiting for a perfect alignment or the wind. His winning tactics secured triumphs and foreshadowed later developments in naval theory that emphasized maneuver warfare and other concepts central to 20th-century naval doctrines.

Ushakov's innovations represent some of the first successful developments of naval tactics that marked a significant departure from the conservative norms of his era. His crews outperformed Ottoman counterparts not simply via numbers but because of their readiness, cohesion, and of course his revolutionary tactics. They became an undefeated combat power whose dynamic style reshaped naval tactics across Europe.

His unorthodox style, which turned conventional sea warfare on its head, allowed him to control key regions like the Ionian Islands without a single defeat, earning him sainthood in the Russian Orthodox Church for his moral and tactical excellence.

When one analyzes Ushakov's tactical genius, one will find it essentially rooted in his systematic rejection of the prevailing "linear template tactics" enshrined in the battle of the line strategy to fight where necessary without the rigid line immobility. He freely adapted tactics to the conditions, never blindly following regulations. He famously echoed Peter the Great's maxim, "regulations should not be followed blindfold because they specify general ways of behavior with no mention of emergency cases and extreme situations."

At the end of his military career in 1804, Admiral Ushakov, recalling his long service in the Russian Navy, wrote: "Thanks be to God, in all the aforementioned battles with the enemy and while the fleet was under my command at sea, through God's will, not a single ship was lost and not a single serviceman was taken prisoner by the enemy." This remarkable record stands unique in naval history. Ushakov created a new paradigm of naval warfare, demonstrating methods and principles of decisive action that would be echoed by later great admirals operating in different theaters.

The tactical evolutions of Ushakov and Horatio Nelson, though operating in different spheres and without direct evidence of mutual influence, exhibit striking similarities. Both were "meleeists" who consciously broke from the rigid line-of-battle doctrine, advocating for decisive action, concentration of force, and aggressive pursuit. Nelson's famous "band of brothers" approach to leadership, which emphasized clear commander's intent and empowered subordinate initiative, resonates strongly with Ushakov's cultivation of initiative and determination among his officers. Perhaps the most direct parallel is Ushakov's inshore attack at Cape Kaliakra, a "violation of the established rules of naval warfare." Ushakov's inshore aggression at Kaliakra prefigured the kind of daring, rule-breaking tactics that Nelson would later employ at the Nile and Trafalgar, suggesting a parallel evolution in tactical thought.

The striking similarities between Ushakov's and Nelson's tactical innovations, particularly their departure from rigid line tactics and their embrace of aggressive, decisive action, suggest a shared recognition of the inherent limitations of the formal prevailing doctrine. The lack of direct evidence that Nelson explicitly learned from Ushakov implies a parallel evolution of tactical thought, and suggests that both commanders, independently or through a shared understanding of fundamental principles, arrived at similar conclusions for how one should achieve decisive victory in the Age of Sail.

This parallel development of universal principles of effective command and combat reinforces the idea that Ushakov's innovations represented a logical and effective progression in naval warfare. It elevates his status from a regional hero to a figure whose tactical genius was on par with, and in some cases preceded, the most celebrated Western naval commanders, in particular Nelson.

While Western historians, as noted by Soviet Admiral Sergei Gorshkov, sometimes "criticize Western European and U.S. naval historians for slighting Ushakov's achievements as being merely those of a pupil of Nelson," some Russian historians contend that Ushakov was ahead of Admiral Nelson by several years in using new fleet tactics. Gorshkov himself pointedly compares Ushakov's capture of the French fortress at Corfu in 1799 with Nelson's inability to take a "weaker" French position on Malta, highlighting a perceived strategic advantage for Ushakov in combined operations.

Fyodor Ushakov was a central figure in the modernization of the Russian Navy in the 18th century. His military operations are still carefully studied in naval academies worldwide today over 200 years later. For his revolutionary tactics and victories, he received:

- Order of St. Vladimir (multiple degrees)
- Order of St. George
- Order of Saint Alexander Nevsky
- Promotion to full Admiral
- Title of Prince of Melfi (honorary)

In 2001, the Russian Orthodox Church canonized him as Saint Fyodor Ushakov, a rare honor for a military leader that recognizes his moral leadership as much as his tactical brilliance. He was subsequently declared the "patron saint of the Russian Navy" and his icon is even carried by some crews of strategic bombers, symbolizing his enduring martial legacy.

His military contributions are also formally enshrined. The prestigious "Order of Ushakov" was established in 1944 during World War II, and is one of its highest naval honors. It is specifically given for "Navy officers who showed outstanding achievement leading to victory over a numerically superior enemy." The establishment of the Order of Ushakov for defeating numerically superior enemies indicates that his aggressive, decisive tactics against overwhelming odds were seen as highly relevant and inspirational for modern Russian naval doctrine.

CHAPTER 17: DON ÁLVARO DE BAZÁN (1526–1588)

Admiral Don Álvaro de Bazán, the 1st Marquess of Santa Cruz (1526–1588), was the architect of Spanish naval supremacy in the Mediterranean and Atlantic during the 16th century zenith of the Habsburg Empire. His victories – from the melee at Lepanto (1571) to the galleon duel at the Azores (1582–83) – cemented his reputation as the indispensable admiral of Spain's golden age.

Bazán understood that naval power must serve broader imperial objectives. True maritime security meant controlling bases, ports, and sea lanes, not just winning battles. This was the real objective for attaining national supremacy, and China is pursuing such a course today. Bazán campaigned relentlessly against corsair bases by blockading North African ports, closing harbors such as Tetouan, and securing Spain's lifelines to the Americas and Flanders. His strategies aimed to destroy the enemy's infrastructure and economic base – a form of strategic warfare that transcended simple fleet actions.

Contrary to common wisdom that prized fleet preservation above all, Bazán favored aggressive action. His audacious, winning strategies became legendary for their consistent results across multiple theaters. Undefeated in battle across a 50-year career, Bazán revolutionized Spanish fleet tactics, and he influenced both ship design and naval warfare for centuries. He was a master of marine infantry tactics, pushing the integration of naval and amphibious warfare to new heights.

His revolutionary strategies – amphibious assaults and combined arms, innovative ship designs, a new pattern for fleet composition, commerce raiding and anti-piracy operations, strategic use of reserves, and leadership infused with chivalric resolve – set him apart from predecessors like the galley-focused Andrea Doria, or peers such as the aggressive Hayreddin Barbarossa. Bazán's approach effectively modernized naval warfare, establishing principles of combined arms and ship design that would dominate the next century. Bazán is often described as the last great galley admiral and the first great sailing admiral – a bridge between two eras.

Galleon Warfare and Ship Design

Bazán was the first commander to successfully demonstrate that galleons, not galleys, would dominate future naval combat. He pioneered the strategic usage of several kinds of ships, refining the design of existing vessels while popularizing galleons as warships, ultimately making possible many successes of the Spanish Empire since the 16th and 17th centuries.

The sixteenth century was a time of experimentation in ship design, rigging, and armament, and these changes brought in their wake new naval designs and tactics. Bazán was one of the outstanding figures of this transitional period, and the only man to win victories as a commander of galleys (Lepanto) and galleons (Azores). The design of the great galleons employed as the backbone to carry the trade between Cádiz and Vera Cruz in New Spain was influenced by both Bazán and Pedro Menéndez de Avilés.

Unlike predecessors reliant on galley ramming and boarding, Bazán emphasized disciplined, massed broadsides where he used heavy cannons to weaken enemies before closing in. He fused galleys with infantry assaults, mounted cannons on masts for fire support, and used ladders for scaling coastal fortifications. Ships became siege platforms and sailors became assault troops, forging a truly integrated naval-infantry force. This was transformed navies from mere transporters to integrated forces, which influenced Spain's imperial expansion and modern amphibious doctrine.

Bazán oversaw the development of larger, more heavily armed galleons with reinforced hulls and higher freeboards, prioritizing both firepower and seaworthiness. This made them ideal for Atlantic operations where galleys could not operate effectively. Basically, Bazán synthesized the best features of Portuguese caravels with the Spanish galleon to produce a new class of ship suitable for transoceanic warfare and amphibious operations.

These ships, with their slimmer hulls, lower profiles, and mixed sails, could weather Atlantic storms and bring overwhelming, sustained cannon power to engagements – marking a decisive break from the oared galleys of his predecessors. They were nimble ships armed with heavier cannon, yet faster and more maneuverable than clunky carracks. They could sustain longer engagements than traditional galley fleets too. By refining the galleon design for warfare, it became the backbone of Spanish naval power for decades, which then made the galleon the capital ship of fleet action.

Bazán's innovation in ship designs, refining galleons and galleasses for versatility, allowed mixed fleets that excelled in firepower and maneuver. He then perfected the usage of galleys, galleasses, and galleons in coordinated formations – creating a flexible force capable of both Mediterranean maneuvers and Atlantic ocean warfare.

Shifting from galley-centric to hybrid fleets of balanced composition supported Spain's global empire by enabling it to counter diverse threats. The heavy guns of galleons provided standoff firepower, while galleasses' oars provided maneuverability for close assaults. By combining the galley's speed with galleon's firepower a combined fleet became more versatile and could adapt to nearly any theater.

In the first engagement between large fleets of carracks and galleons fighting at a great distance from the mainland, at the Battle of Terceira (also known as the Battle of Ponta Delgada) the mercenary fleet of French, English, Dutch and Portuguese ships under Filippo di Piero Strozzi were severely defeated by a squadron under Bazán. The battle, fought in the open Atlantic, demonstrated the decisive superiority of broadside-armed galleons over mixed fleets that still relied on galley-style tactics.

Bazán won by adapting galley-style tactics to his galleon fleet. His galleons, heavily armed with cannon firepower, were deployed in a line-abreast formation that mimicked the tactical style of galley fleets while he was deploying the superior range and durability of galleons. The heavier cannon fire of the larger Spanish galleons proved decisive.

This adaptation of galley formation principles (tight formations and coordinated movements) to galleon warfare created a new paradigm for Atlantic naval combat, maximizing the firepower of multiple gun decks while maintaining tactical coordination. The Spanish victory demonstrated the superiority of galleons over mixed fleets and led directly to Spain's confidence in launching the Armada against England six years later. In essence, the battle showcased the decline of galley dominance and the rise of galleons as the future of naval warfare.

Integration of Galleys and Galleons into a Hybrid Fleet

Bazán gradually became a master of mixed fleet tactics where he integrated galleys, galleons, caravels, and small craft for multi-role operations. He understood that a properly balanced fleet could exploit both tactical agility and heavy firepower because galleons could carry heavy artillery and endure in the open sea while galleons had speed and maneuverability for fighting in calm or confined waters. He placed galleons in the main line of battle to deliver sustained broadsides and absorb punishment, and galleys on the flanks or front to maneuver aggressively, disrupt enemy formations, and deliver rapid, close-quarters fire. His galleons would deliver crushing broadsides while his galleys would cannonade the enemy, forcing them to halt or turn.

Typically sailing ships formed the backbone of his fleet while galleys acted as tactical strikers that allowed him to shape the tempo of the engagement. Bazán's flexibility in fleet composition allowed him to respond to changing battle circumstances, reinforce weak points in formations, or concentrate on enemy vulnerabilities. He was not slowed by rigid orthodoxy; rather, he pioneered ad hoc battle groups that combined various ship types and functions to suit each mission.

As an example, in the campaign against Ottoman corsairs at Tetuan (1564–65), Bazán formed a composite squadron of galleys, brigantines, caravels, and chalupas. He deployed fast brigantines for reconnaissance, chalupas for amphibious assaults, and delivered overwhelming gunnery support from galleys at critical moments – a clear demonstration of combined-arms warfare.

Other successes that demonstrated his hybrid fleet composition were at the Battle of Lepanto where Bazán's galleons supported galley reserves, sinking Ottoman ships. Bazán's galleasses, with 40 guns each, shattered the Ottoman lines. At the Battle of the Azores, galleons bombarded the enemy while Bazán's galleys flanked, capturing 10 ships. At Peñón de Vélez, Bazán's hybrid fleet landed troops under galleon cover.

Defensive Withdrawals: Preservation as Power

Unlike many aggressive commanders of his day interested in glory, Bazán was like many great commanders who prioritized the long-term survival of his navy (valuing sustainability) over fame. Bazán's strategy of defensive withdrawals, where he would retreat to avoid losses, protected his fleet from destruction and preserved it for future battles, which made the withdrawals victories in disguise. Withdrawals, in fact, sometimes lured enemies into overextension, which allowed for counterstrikes. One of the reasons that Bazán never lost a ship was because of his strategy of withdrawals.

As examples,

- At Preveza, as second-in-command, Bazán advised withdrawal, saving the fleet from Barbarossa.
- At Muros Bay (1543), Bazán withdrew after initial gains, preserving his ships.
- At Malta Relief (1565), he withdrew after resupply, avoiding an Ottoman trap.

Battle Strategy

Unlike earlier admirals who accepted set-piece, head-to-head engagements, Bazán emphasized maneuver – speed, positional advantage, and the concentration of fire at the decisive point. He often seized the initiative by closing swiftly with the enemy, using wind and weather to maximum effect.

For instance, at the Battle of Muros Bay, faced with a larger French fleet Bazán leveraged the superior size and firepower of his galleons. Bazán believed in decapitating the enemy command, as did the Admirals Ushakov and Nelson after him, so he attacked the French flagship directly by using his own ship as a battering ram, and sank it. This triggered French disorder and total victory because afterwards, during the ensuring confusion and panic, his smaller fleet captured or sank almost the entire French armada – an outcome that awed Europe. This approach of "strike at the enemy's head" became his hallmark tactic.

Several examples of his fluidity in battle strategies:

- At Ponta Delgada, he adopted a flexible line-abreast formation – like a galley fleet – and closed rapidly into the enemy despite numerical balance. His heavier cannon fire overwhelmed Strozzi's mixed privateer force.
- As commander of the Holy League's rear guard at Lepanto, Bazán adeptly redirected his reserve division to shore up weak points and preserve allied cohesion – turning the tide at critical moments.
- Bazán spearheaded amphibious operations such as the relief of Malta in 1565 and the capture of the Azores in 1582. His forces combined naval bombardment with strategic landings to secure coastal strongholds.
- Bazán's quick decision-making – such as sailing rapidly to Muros to surprise the French fleet – enabled victories despite being often outnumbered. He rarely delayed battle over logistics.

Bazán was also known to frequently employ a modified crescent formation to achieve fleet encirclement, where his galleys and galleons formed a concave line, drawing enemies into a killing zone before counterattacking. At Lepanto, this very tactic allowed the Holy League to trap and destroy the Ottoman fleet. His squadron broke the Ottoman left wing by focusing fire on their flagship, causing a chain rout.

Naval and Amphibious Warfare

Another tactical uniqueness that separated Bazán from admirals who confined their battles to open water, was his mastery of combining amphibious assaults with naval warfare where he coordinated galleys with troop landings to seize coastal strongholds. His development of amphibious operations was decades ahead of his contemporaries.

As he was one of the historical innovators of using infantry troops to take beaches or lands, one might even call Bazán the father of marine infantry for he was decades ahead of his contemporaries. Having marine infantry transformed navies from mere transporters to integrated forces, and this revolutionary tactic influenced Spain's imperial expansion while shaping its modern amphibious doctrine.

Unlike predecessors focused primarily on galley ramming, Bazán refined vessel designs by mounting cannons and ladders on masts so that they could make direct assaults on fortifications. Amphibious assaults allowed Bazán to strike at enemy ports and fortresses while minimizing naval risks. By employing precise timing and coordination he could overwhelm defenders and secure strategic bases with fewer casualties.

Bazán commonly coordinated naval bombardment, troop landings, and logistical support into a single devastating strike. His most famous application of such combined operations was once again at the Battle of Ponta Delgada, where Bazán's galleons supported landings in the Azores. There he annihilated a Franco-Portuguese fleet of 60 ships with just 28 and then stormed the Azores, securing the islands for Spain and Spanish control of the Atlantic.

During the conquest of Peñón de Vélez de la Gomera (1564), Bazán's formed a composite squadron of galleys, brigantines, caravels, and chalupas using each for different functions, and delivered overwhelming gunnery support from galleys at critical moments. His 90 galleys landed 4,000 troops under cannon cover to capture the Moroccan rock fortress from Barbary pirates despite heavy resistance. His mast-mounted guns provided fire support for the attack, turning a naval expedition into a land victory. His combined arms approach thereby overwhelmed the garrison, securing Spain's North African foothold.

During the Conquest of Tunis (1573), Bazán commanded 100 galleys and landed 20,000 troops against Ottoman defenses, capturing the city and 42 ships. His integration of naval artillery with infantry assaults once again routed the defenders.

At the Coron Siege (1532), as second-in-command, Bazán's galleys provided fire support for landings, capturing the Ottoman-held Greek port. During the relief of Malta, Bazán's fleet landed reinforcements under fire, breaking the Ottoman blockade and saving the Knights of St. John. In effect, Bazán's landings saved Christendom in the Mediterranean and the Mediterranean World.

Commerce Raiding and Anti-Piracy to Establish Economic Dominance

Bazán's commerce raiding and anti-piracy operations were a systematic, state-run campaign that strangled French, British, Dutch and Barbary corsair shipping in order and make the Spanish treasure route almost untouchable for three decades. He focused on disrupting the shipping logistics of his enemies as a type of economic warfare that would disrupt their economies. Raiding served several purposes: it drained enemies of resources, disrupted their naval funding, forced them to spend more on protection than expansion, and securing Spanish trade routes. At the same time, Bazán's naval patrols reduced piracy in order to secure his own seas.

As examples, at Peñón de Vélez Bazán raided Barbary bases, capturing ships and crippling the pirates. He raided Moorish supply lines at Alpujarras (1569), an act of economic strangulation that crushed rebellion.

Strategic Reserve Deployment

At the Battle of Lepanto, Bazán demonstrated a mastery of reserve forces – one of the most challenging aspects of naval command.

Behind the galleasses (employed to spread preliminary confusion), the Christian fleet advanced in four squadrons. Don Juan commanded the center; the Venetian, Agostino Barbarigo, the left; Philip's admiral, Giovanni Andrea Doria, the right; and Bazán commanded the Holy League's reserve squadron of thirty-five galleys. The coalition fleet consisted of 208 galleys against 230 Ottoman vessels. Here, as commander of the reserve fleet, Bazán displayed excellent seamanship and played an important role of decisive intervention in the crushing of the Turkish fleet.

With strategic restraint, Don Alvaro de Bazán had carefully kept his ships out of the fray until the moment came when he was most needed. Late in the battle, the Christian victory was almost averted when Uluch Ali, by threatening to outflank Doria's squadron, drew it toward the open sea and then penetrated the gap that was thus formed. The Maltese force bore the brunt of the attack and suffered tremendous casualties.

Disaster loomed near, but was averted by Bazán who committed the Christian Reserve at precisely the right time. Entering the fight, he rescued the center of the Holy League from the Turkish vessels that had surrounded them before turning his squadron south to aid the outmanned Doria.

Bazán's Legacy

Over a career spanning five decades, Bazán fought in over 40 naval engagements – from the Barbary Coast to the Atlantic – and never lost a single battle. He understood that naval supremacy was the backbone of the Spanish empire, and by seizing the Azores his victory at Terceira ensured Spanish control over Atlantic trade routes. Though he died before its launch, Bazán was also the original architect of the Spanish Armada; his 1583 letter to Philip II laid out the plan to invade England.

Bazán influenced ship design for centuries by refinement of the galleon in a way that altered the strategic calculus for both Atlantic and Mediterranean powers. His combined-arms operations – using ships of varying size, capability, and purpose as an effective whole – became the template for future sea power. He instituted flexible formations to defeat rigid ones and tailored strategies to each battle to deploy and redeploy as needed, improvising in chaos, would made him unbeatable. Overall, his various innovations in superior ship design for battlefield dominance and fleet composition, together with his winning tactics, secured Spanish naval dominance and shaped the Age of Sail.

Bazán was a father of modern amphibious warfare and marine infantry tactics who blended naval, amphibious, and logistical operations in ways rare for his age: capturing islands, reinforcing armies from sea, and blockading rivers – each enhancing Spain's strategic reach. He proved that integration of naval and land forces wins wars, and that decapitation strikes helped to win battles by sowing confusing and breaking enemy morale.

His mastery of strategy and tactics against superior foes secures his title as one of history's great naval commanders. His tactics revolutionized naval warfare, setting him apart as an innovative and undefeated commander who shaped the Age of Sail.

CHAPTER 18:
MY TOP 5 NAVAL COMMANDERS

One might partition these fifteen top naval commanders into various categories such as "Best of the Best," "Most Decisive Single Victory," or "Biggest Impact on Naval Warfare," but I prefer using these three groups:

- **Unmatched Legends** (like Horatio Nelson and Yi Sun-sin)
- **Great Strategists** (like Fyodor Ushakov, Chester Nimitz and Michiel du Ruyter), and
- **Fearless Raiders and Tacticians** who were exceptional in some areas but lacked the scale or consistency of the Unmatched Legends (such as Francis Drake, Andrea Doria and Thomas Cochrane).

In reviewing their histories, we find in these super naval commanders a strong tendency towards aggressiveness, which is a trait also found in the most successful generals of all time. Several repeating themes responsible for their greatness (in no particular order) have also consistently appeared in these tactical biographies:

- Breaking with the traditional fighting tactics of the time
- Using revolutionary ship designs
- Decentralized command that trusts subordinates to make independent decisions
- Adaptive fleet formations; unique formation strategies

- Crews capable of executing complex maneuvers due to extensive training that made them more efficient than opponents
- Having and properly using a reserve fleet
- Adopting new technologies first or fast; applying continuous innovation
- Artillery speed and precision firing from training, and unique bombardment strategies
- Integrated naval and land forces (amphibious operations)
- Usage of intelligence networks, and knowledge of coastal current and tidal patterns
- Precision ambushes
- Flagship targeting
- Pursuing total enemy annihilation or ship capture
- Calculated risk
- Stealth, deception, and psychological warfare

My Top 5 Naval Commanders

Now for the most controversial part – who would I choose as the top 5 commanders of all time? I certainly have a short list of favorites whom I consider the best naval commanders in history, particularly due to their win-loss battle record, revolutionary tactics, and lasting mark on transforming naval warfare.

My selection of the five greatest naval commanders of all time actually becomes a list of six, with a tie between two admirals for inclusion. My candidates for the top tier of greatest maritime fighters in history is here presented in order of their win-loss record:

- Fyodor Ushakov (43-0 win-loss record)
- Don Álvaro de Bazán (40-0 win-loss record)
- Yi Sun-sin (23-0 win-loss record)
- Horatio Nelson (8-1 win-loss record)
- Tie: Thomas Cochrane (5-0 win-loss record) or Hayreddin Barbarossa (5-1 win-loss record).

Whether these naval commanders (all of whom were admirals) should be included in the top tier or not, when we examine their track records it becomes apparent that these men and their tactics merit further study for the topic of naval warfare.

What we have not yet covered are the naval strategies of Russia and China, as well as future adjustments the U.S. Navy must ponder because of the invention of AI, inexpensive drones, autonomous unmanned vessels, hypersonic missiles, cyber warfare and so forth. That will wrap up the final chapter of the book.

CHAPTER 19: RUSSIA, CHINA, AND CONCERNS GOING FORWARD

To comprehend the future of the U.S. Navy and its potential responses to the rise of global powers like China and Russia, a deep understanding of how a country develops national prosperity is essential. In previous eras this was greatly related to commodity exports and resource extraction, but the manufacturing era has changed this altogether.

Navies have historically protected and sustained the sources of a nation's wealth because trade was transported via the oceans, and so for the unbroken flow of commerce those trading routes needed protection from piracy and blockades. This chapter challenges conventional wisdom with a better understanding of how nations achieve national prosperity along and global influence, and how navies are used to protect this status. This leads to important implications as to how to structure the American navy in composition and function for the times ahead.

A navy is more than a military force for defending coastlines and deterring adversaries. We have seen that it is an instrument of national policy that actively supports a nation's pursuit of abundance and prosperity. However, to be effective a navy must adapt its form and function to the prevailing economic realities and fiscal limitations of its nation. When these change, a navy must adapt. While American naval strategy has been heavily influenced by the history of the British Empire and its methods for developing and protecting its trade routes, a more robust framework for understanding national prosperity is needed.

Reinert's Thesis of Industrialization as the Engine of Wealth

The most insightful analysis for naval students seeking to understand the foundations of national wealth comes from Erik S. Reinert's seminal work, *How Rich Countries Got Rich ... and Why Poor Countries Stay Poor.* Reinert's central thesis is that a nation's prosperity hinges on its economic structure, specifically whether it prioritizes industrial activities or relies on a commodity-based economy.

Reinert inarguably proves that most modern wealthy nations achieved their prosperity through industrialization, which was fostered by deliberate policies such as tariffs and subsidies that protected state industries. After you build up your industrial competitiveness, only then can you talk about free trade in order to sell more of your goods abroad. In contrast to manufacturing nations that have become rich, countries that remain trapped in poverty have typically relied on the export of raw materials. This distinction between industrial and commodity-based economies is crucial for understanding the dynamics of global trade and economic power.

Reinhart views the wealth of nations as dependent upon whether they prioritize industrial activities or the export of natural resources. Economies based on the export of raw materials, such as minerals or commodities, face inherent disadvantages that inevitably lead to stagnation and decline just as Spain saw with its dwindling silver imports from the Americas. These economies are susceptible to volatile prices because they compete with other similar commodity producers on the same world markets, and they experience diminishing returns over time as they exhaust commodity resources.

For example, a nation that becomes a major oil exporter may enjoy a period of increasing wealth, but this prosperity is vulnerable to fluctuations in global oil prices and the discovery of alternative energy sources. Furthermore, as the most accessible oil reserves are depleted, the cost of extraction increases, eroding oil profit margins over time. Essentially, industrial economies grow rich over time for many reasons while commodity-based economies eventually become poorer.

The principle of diminishing returns – decreasing profits – applies to nearly all commodity-based industries. In mining, the highest-grade ores are extracted first, and subsequent efforts require more effort and monies to reach lower-grade deposits. Therefore marginal production costs increase over time as resources are depleted. Similarly, in agriculture, the most fertile land is cultivated first, and expansion necessitates the use of more fertilizer on less productive land which leads to higher marginal costs. When agricultural commodities are sold on the global market, competition from lower-cost producers further squeezes profits.

In contrast, industrial economies benefit from a virtuous cycle of increasing returns as more is produced. As manufacturing volume increases, production costs decrease due to learning curves (efficiencies) and innovation. This allows for both higher profit margins as sales volume increases if the product has differentiated itself from competitors to stand out. Manufactured goods can be differentiated through unique features, enabling producers to command premium prices for their own brand that can avoid direct price competition.

By analyzing the history of how countries became rich or poor, Reinert has consistently found that economic development requires the nurturing of increasing-return activities (e.g., manufacturing, innovation) rather than decreasing-return activities (e.g., agriculture, resource extraction, fishing, timber). To become rich and stay rich today, a country must become a manufacturing exporter of unique products. If one argues that an information-based or service-based economy, rather than a manufacturing-based economy producing real items, can be the basis of wealth-making, one must note that such activities can be easily duplicated across the world resulting in countless competitors. India, for instance, quickly became a leading software developer that partially knocked the U.S. off its pedestal, and now other countries do the same.

These principles of economic prosperity are not a recent discovery. As early as the 17th century, astute observers recognized that the path to wealth lay in value-added production rather than the mere export of raw materials (whose shipments navies have protected for centuries). A prime example of this early understanding is Philipp von Hornigk's "Nine Points on How to Emulate the Rich Countries (1684)," as cited by Reinert.[1] These principles, summarized below, offer a timeless guide to the methods of wealth creation and economic development:

---

[1] *How Rich Countries Got Rich … And Why Poor Countries Stay Poor*, Erik Reinert, (Carroll & Graf Publishers, New York, 2007), pp. 313-315).

First, to inspect the country's soil with the greatest care, and not to leave the agricultural possibilities of a single corner or clod of earth unconsidered. Every useful form of *plant* under the sun should be experimented with, to see whether it is adapted to the country, for the distance or nearness of the sun is not all that counts. Above all, no trouble or expense should be spared to discover gold and silver.

Second, all commodities found in a country, which cannot be used in their natural state, should be worked up within the country; since the payment for *manufacturing* generally exceeds the value of the raw material by two, three, ten, twenty, and even a hundred-fold, and the neglect of this is an abomination to prudent managers.

Third, for carrying out the above two rule, there will be need of people, both for producing and cultivating the raw materials and for working them up. Therefore, attention should be given to the population, that it may be as large as the country can support, this being a well-ordered state's most important concern, but, unfortunately, one that is often neglected. And the people should be turned by all possible means from idleness to remunerative *professions*; instructed and encouraged in all kinds of *inventions*, arts and trades; and, if necessary, instructors should be brought in from foreign countries for this.

Fourth, gold and silver once in the country, whether from its own mines or obtained by *industry* from foreign countries, are under no circumstances to be taken out for any purpose, so far as possible, or be allowed to be buried in chests or coffers, but must always remain in *circulation*; nor should much be permitted in uses where they are at once *destroyed* and cannot be utilized again. For under these conditions, it will be impossible for a country that has once acquired a considerable supply of cash, especially one that possesses gold and silver mines, ever to sink into poverty; indeed, it is impossible that it should not continually increase in wealth and property. Therefore,

Fifth, the inhabitants of the country should make every effort to get along with their domestic products, to confine their luxury to these alone, and to do without foreign products as far as possible (except where great need leaves no alternative, or if not need, widespread, unavoidable abuse, of which the Indian spices are an example). And so on,

Sixth, in case the said purchases were indispensable because of necessity or *irremediable* abuse, they should be obtained from these foreigners at first hand, so far as possible, and not for gold or silver, but in exchange for other domestic wares.

Seventh, such foreign commodities should in this case be imported in unfinished form, and worked up within the country, thus earning the wages of *manufacturing there.*

Eighth, opportunities should be sought night and day for selling the country's superfluous goods to these foreigners in manufactured form, so far as this is necessary, and for gold and silver; and to this end, *consumption*, so to speak, must be sought in the farthest ends of the earth, and developed in every possible way.

Ninth, except for important considerations, no importation should be allowed under any circumstances of commodities of which there is a sufficient supply of suitable quality at home; and in this matter neither sympathy nor compassion should be shown foreigners, be they friends, kinsfolk, *allies* or enemies. For all friendship ceases, when it involves my weakness and ruin. And this holds good, even if the domestic commodities are of poorer quality, or even higher priced. For it would be better to pay for an article two dollars which remains in the country than only which goes out, however strange this may seem to the ill-informed.

These points clearly demonstrate that transforming raw materials into higher value-added goods (manufacturing) has long been recognized as the key to national wealth. This is one reason why Alexander Hamilton, in his

Report on Manufacturers of 1790, expressed opposition to American nationals only serving as a raw material exporter to Britain.

In short, prioritize the development of domestic resources, promote domestic manufacturing where the resultant value of manufacturing far exceeds the value of the raw material, cultivate a skilled population, internally retain gold and silver for circulation to ensure a continuous increase in wealth, encourage consumption of domestic products rather than foreign goods, barter for essential imports rather than purchase them with gold or silver, import foreign commodities in unfinished form and process them within the country to capture the value of manufacturing, export manufactured products (for gold and silver), and protect domestic industries to preference domestic strength over foreign friendships.

Manufacturing is therefore the way to go, which was recognized as early as the 1600s because it quickly results in lower production costs with a higher potential sales volume (when you produce a quality product for which there is growing demand). History clearly shows that countries which industrialized (e.g., Britain, U.S., Germany, and later East Asian nations) achieved their tremendous wealth by moving into manufacturing and value-added production. Commodity-dependent economies (reliant on raw material exports) tend to face declining terms of trade, price volatility, and lower productivity growth, keeping them poor over time.

Therefore, you don't want "export commodities" as your national geopolitical strategy even if this is your "comparative advantage." This is at best a short-term revenue strategy. To become wealthy, you want your country processing resource raw materials into higher value-added goods you can export so that you achieve higher margins and greater profits. You want to focus on manufacturing where there are higher margins over time, due to the benefits of the learning curve for cutting costs, and where there is more demand for your products as populations grow. This is a formula for wealth, including national wealth. Wealth = (Increasing margins) * (Increasing sales volume).

History also shows that nearly all rich countries, including Great Britain, the United States, and Germany, used protectionist policies (tariffs, subsidies, and state intervention) to protect and nurture their domestic industries during their early stages of development. Reinert demonstrates that Great Britain, a pioneer of industrialization, built its manufacturing base behind a wall of protectionist policies, such as the Navigation Acts.

However, once it had established its industrial dominance, only then did Britain advocate for free trade to prevent other nations from following the same path and becoming competitors. Typically it has advocated for policies of deliberately controlled economic disintegration in other countries – a type of imperial sovereignty over them – that will restrict their industrial output and keep them as customers of British goods.

Successful nations in history have copied the industrial policies and development pattern of richer countries (e.g., the U.S. copying Britain, Japan copying the West) while ignoring misleading rhetoric. Free trade was only adopted *after* achieving industrial dominance, not during the catch-up phase. Poor countries stay poor because colonialism, and later free-trade policies (e.g., structural adjustment programs), forced poor nations into commodity specialization, locking them into low-growth traps. Reinert critiques mainstream economics (like comparative advantage theory) for ignoring the historical reality that industrial diversification, *not raw material resources*, drives long-term wealth. You certainly need access to strategic resources and raw materials but processing them in a way that adds value brings the wealth that nations seek. Of course, access is the first concern; wars are fought for those resources (or water and food) when they are in short supply and you cannot gain them through treaty or buy them on an open market.

Nations who understand these principles are always trying to mislead other nations in order to keep them as customers for their own goods, and to prevent them from becoming competitors. Britain actively opposed U.S. tariffs in the 19th century because a commodity-exporting America would have served British industrial interests best. An industrialized United States, on the other hand, would have become a formidable competitor and cut into its profits. Reinert highlights how Britain, a pioneer of industrialization, used protectionist policies to build its own manufacturing base, but Britain later advocated free trade to keep other nations dependent on commodity exports as their road to revenue. Britain is notorious for giving bad advice on purpose, as if it was guided by some secret desire to dominate the planet due to a "reign or die" mentality. One of its greatest fears was that the principles of the American economic system would spread and create strong, prosperous sovereign nations that were immune to imperial policies of economic backwardness, financial control and depopulation.

Specifically, Britain attempted to prevent the United States from adopting tariffs and industrializing in the 19th century, understanding that industrialization would make the U.S. rich, while a commodity-based economy (e.g., exporting cotton, tobacco, fish, coal or timber) would keep it subordinate and poor. The U.S., however, under the guidance of figures like Alexander Hamilton, defied British pressure and implemented tariffs to protect its fledgling industries, paving the way for its own economic ascent. This historical example underscores the hypocrisy of powerful nations that advocate for free trade while having used protectionism to achieve their own wealth.

Reinert argues that the global trade rules and institutions of the modern era, which have been shaped by wealthy nations, often discourage industrialization in developing countries, effectively locking them into a state of commodity dependence and poverty. He calls for these nations to adopt strategic industrial policies, emulating the historical path of today's rich countries, to escape the poverty trap.

In the modern era, many have claimed that a nation can achieve wealth through financialization by becoming a financial center or selling financial products. However, these are paper games that cannot lead to lasting wealth. While some cities have temporarily grown rich by establishing themselves as banking centers or tax havens, their specialization made them vulnerable.

After reviewing the economic history of cities and nations that have become rich over the past two millennium, international investor and economic historian Marc Faber (*Riding the Millennial Storm*) noted, "Financial centers simply do not control their own destiny. They are controlled by their debtors. Without a powerful military, they can, in the worst case, never enforce their claims." This is a warning that financial centers like London and Dubai, which do not have a strong manufacturing base, should heed. A stable economy depends on production rather than on passing financial papers back and forth. A nation that cannot manufacture or market and ship its products loses its strength for its economic security is its national security.

Similarly, the new utopian vision of wealth through the digitization of information is also a mirage of stability. While the digital revolution has created economic booms, the underlying processes can be replicated anywhere in the world, quickly eroding the initial advantage of any productivity bump that makes anyone a competitive supplier. Ultimately, people need tangible products to live, and economies built solely on ephemeral services or information are unsustainable. There will always be economic booms as fascination with the "newest shiny thing" reaches an extreme, but after the major profits have been made and people start deriving better estimates of the future, bust (decline) is an inevitable outcome.

Nonetheless, the secret to riches from economic development requires nurturing increasing-return activities (e.g., manufacturing, innovation) over decreasing-return activities (e.g., agriculture, resource extraction). In ancient times shipping goods from one region to another brought wealth because of pricing arbitrage, but today the formula is to generate increasing returns to scale from increasing sales while benefitting from lower costs due to the steady quality of goods, increasing volumes and learning curve innovations. Increasing demand for one's goods multiplied by an increasing margin equals wealth.

In conclusion, the long-term prosperity and security of a nation are inextricably linked to its economic structure. The historical record, as elucidated by Erik Reinert and others, demonstrates that industrialization is the most reliable path to sustained wealth and global influence. Wealth supports a large military whereas poor countries cannot afford one. Commodity-based economies, with their inherent vulnerabilities, cannot provide the stable foundation required for enduring national power.

Now, how is this all relevant?

For the U.S. Navy, this understanding is of paramount importance. A navy's primary mission is to protect the nation's economic lifelines, and in the 21st century, this means safeguarding the sources of industrial strength. As we have seen, many naval strategies in the past were designed to protect a nation's trade flows. As the United States confronts a world of rising powers and increasing geopolitical competition, it must recognize that its military might well become meaningless without a robust and innovative industrial base producing actual physical products. The lessons of history are clear: to remain a global leader, a nation must be a maker of goods, not just a consumer or trader of goods.

History also shows that nearly all rich countries, including Great Britain, the United States, and Germany, used protectionist policies to nurture their domestic industries during their early stages of development.

Reinert clearly shows that Great Britain, a pioneer of industrialization, built up its manufacturing base behind a wall of protectionist policies. However, once it had established its industrial dominance, Britain advocated for free trade to prevent other nations from following the same path.

For instance, Britain actively opposed U.S. tariffs in the 19th century because a commodity-exporting America is what would have served British industrial interests; an industrialized United States would have become a formidable competitor who decreased its exports. The U.S., however, defied British pressure and implemented tariffs to protect its fledgling industries, paving the way for its own economic ascent.

This historical example underscores the hypocrisy of powerful nations that advocate for free trade while having used protectionism to achieve their own wealth, or which advise developing nations to further develop their export of global commodities rather than turn to industrialization. It is sometimes said that Britain still acts in similar misdirecting ways with the intent to pull the U.S. back under its influence as in colonial times, using its intelligence agencies to sprout tentacles and launch operations involving various American officials and even American elections. "Allies" are not supposed to do this. Nevertheless, history shows Britain is an expert at long-term subversion to bring other nations under its control, and a known master of "color revolutions," which are fake insurrections that disguise foreign-sponsored coups as a people's uprising.

As stated, Reinert argues that the global trade rules and institutions of the modern era, which have been shaped by wealthy nations, often discourage industrialization in developing countries, effectively locking them into a state of commodity dependence. He calls for these nations to adopt strategic industrial policies, emulating the historical path of today's rich countries, to escape the poverty trap.

Innovation Cycles: The True Engine of National Prosperity

The real force powering the wealth of nations is not commodities or even industrialization, but large-scale, multi-decade innovation waves that sweep economies with new products, trade route discoveries, and economic efficiencies. While commodity booms can produce economic leaps forward, such prosperity booms typically fall within the framework of these larger innovation waves.

Russian economist Nikolai Kondratieff identified famous prosperity cycles of 45-60 years (averaging 57 years) that capture the economic effects of such innovations, which became known as Kondratieff Waves or K-waves. Originally based on agricultural price cycles, Kondratieff waves have evolved to become more technology based as manufacturing industry has become the primary basis of modern economies. Now we can consider them waves of innovation, or waves of technological progress that produce economic gains in nations because they introduce new materials or methods, lower costs, increase productivity, or make selling easier in some way. By increasing productivity in some way they cut costs dramatically, freeing money for other productive uses in economies. Sometimes they reflect grand political changes that bring peace instead of war (which lowers the cost of business) or lower taxes, which thereby creates economic booms as well. Most analysts refer to these as sixty-year cycles, a convenient round number that coincides with approximately three generations.

The Kondratieff wave has been traced back in price series and other economic statistics more than 1,000 years. Richard Mogey and others, for instance, have found long waves in prices back to the 10th century. George Modelski and William Thompson (*Leading Sectors and World Powers*) traced K-waves back to Sung dynasty China, approximately 1,000 years ago. Michael Alexander *(The Kondratiev Cycle)* also traced the leading economic sectors of K-waves back to the early 1500s by studying the S-shaped diffusion curves of leading industries that powered economies.

The Kondratieff waves are characterized by alternating periods of high growth (upswings) and stagnation or recession (downswings). In modern interpretations, particularly those influenced by Joseph Schumpeter's "creative destruction" theory, these waves are fundamentally driven by clusters of technological innovations rather than purely economic factors like capital investment or commodity prices.

Innovations bunch together during downswings, leading to revolutionary breakthroughs that launch new leading economic sectors, boost productivity, and propel the next upswing. This creates an S-curve pattern: irruption (emergence of innovations), frenzy (rapid adoption and speculation), synergy (maturation and widespread integration), and maturity (saturation, leading to the next cycle). Each wave transforms industries, skills, and economies, built around a scalable "technology core" of innovative change that disrupts established systems but enables economic rejuvenation.

If you can master the Kondratieff wave, which spans decades, you can harness the most powerful prosperity forces within a nation. The key to greatest benefit is invention of the relevant innovation or its early adoption – tapping into the new wave as early as possible because the early bird gets the biggest advantage. Kondratieff innovation waves are directly responsible for the rise and fall of economic fortunes in many nations over time, and have directly affected the historical fortunes and functions of navies. Kondratieff cycles even explain the pattern of interstate conflict where countries go to war with one another.

Michael Alexander's 2002 book, *The Kondratieff Cycle*, expands on his prior work analyzing stock market cycles, applying analysis to over seven centuries of Western social, political, and economic history. His core argument is that Kondratiev cycles are not isolated economic phenomena but are deeply intertwined with social and political cycles. Generational shifts influence and are influenced by Kondratieff waves, creating a synthesis that also explains broader societal trends. Economic upswings align with civic generations focused on institution-building, while downswings correlate with reactive generations and periods of crisis.

The Kondratieff wave is a very long economic wave produced by groundbreaking innovations and inventions, such as the automobile or printing press, that launch technological revolutions, which in turn power economies along an S-growth curve pattern for a certain length of time until a new one emerges. These innovations are the primary triggering force behind new leading industries and industrial sectors that produce most of the economic growth within a country for a sustained period. By riding atop a Kondratieff wave and engaging with it fully, a nation can create one of the most productive economies in the world and gain competitive advantages over contenders. When a country ignores these innovations, however, it falls behind. A nation's fortunes rise or fall based upon its ability to lead or adapt to each new innovation cycle.

The automobile innovation wave – whose strongest up move section occurred between the 1920s and 1950s – illustrates how these cycles work. The invention of the automobile produced two large economic sectors in adoptive nations – the automotive industry and the petrochemical industry. For approximately half a century, these industries and their subsidiary sectors produced the greatest investment returns while employing the most people in any nations that embraced them. Commercialization of the automobile resulted not just in auto sales and petrochemical development but in massive nationwide construction projects for highways, bridges, and roads, thus creating permanent infrastructure that benefited participating economies. The automobile and oil-gas industry played a leading function in economies, directing primary economic growth for decades.

Steel and tire manufacturers, oil and gas heating system manufacturers, fuel power stations, gas stations, banks, insurance companies, tourism companies, and numerous suppliers of metal, plastic, and electronic parts all benefited from this innovation wave. An enormous network of suppliers, customers, and users evolved across the globe, creating tremendous prosperity.

The automobile and petrochemical industry generated many consumer, social, and economic benefits and also created new professions. Trucks made deliveries faster, greatly expanding market areas that businesses could serve. A new era of mass transit was born as cars gave people high personal mobility, changing society on a grand scale. In short, the automobile-petroleum innovation wave changed virtually all of society in the direction of greater prosperity.

The wave prior to the automobile era, from about 1870 to 1890, was powered by scientific discoveries in electrotechnology and chemical processes. Electrical communications started spanning the globe, electric motors replaced human muscle power, electrical lights became popular, and material living conditions improved dramatically. This became an era of Mass Consumption where many people received higher education and increased in importance as voters and consumers.

The innovation wave prior to the era of Mass Consumption, starting around 1830-1850, was the Age of Steel, Electricity and Heavy Engineering. This wave arose due to the invention of the Bessemer furnace, which allowed steel of superior strength and quality to be produced, enabling entirely new products. The steel industry became the leading industry defining the new economy, although tremendous growth was also seen in railroad, iron, coal, and chemical industries, many connected with steel innovations.

Steel was the major material component of railways, ships, machines, construction equipment, weapons, and bridges that produced an industrialization boom. Stronger steel rails made it possible for trains to carry heavier loads, reducing transportation costs by a factor of 200. Factory workers became the largest occupational group in society, leading to the formation of unions and socialist groups that dramatically changed human culture.

Prior to the steel wave was the Age of Steam, triggered by the invention of the steam engine (~1770). The textile industry was the largest beneficiary because more textiles could be produced quicker and at lower cost than by using human labor. Textiles, especially cotton, became the leading, most prosperous industry. The production of fabrics, yarns and clothing shifted from home-based to factory-based manufacturing, causing mass migration from the countryside to cities. Cities upgraded by building new housing, streets, water systems, and sewage systems. Factory workers arose as a new social class and began fighting for political rights.

Economist Carlota Perez has provided a complementary framework for understanding these innovation cycles in her work, *Technological Revolutions and Financial Capital*, where she describes in depth five major Kondratieff wave technological revolutions since the Industrial Revolution. Each revolution has brought a new "techno-economic paradigm" – a new common-sense "best practice" model for innovation, production, and organization for the economy. If we restrict ourselves to just these last several technology innovation waves as our pattern, it can be seen that each technological revolution spreads through the economy in several distinct phases.

For this K-wave cycle, a new discovery or technology first emerges with astoundingly large potential effect on the economy. Financial capital (venture capitalists, speculators) recognizes its potential and floods in, funding a frenzy of experimentation and startup creation. Success in this first phase leads to wild speculation, inflated asset prices, and a massive bubble (e.g., Railway Mania, the Roaring Twenties, the Dot-com Bubble) that inevitably collapses. Society then creates an institutional framework (new regulations, education systems, infrastructure) to guide the technology for broad prosperity, and the focus then shifts to widespread deployment of the new technology. This phase is characterized by growth, widespread prosperity, and the full integration of the new paradigm across all industries (e.g., the Victorian Boom, Post-WWII boom). The technological revolution eventually runs out of steam as it fully permeates society in S-curve growth;

growth reaches a natural limit. Markets become saturated, innovation slows, and production capital seeks higher returns in financial speculation, setting the stage for the next revolution.

Typically, some new type of expansive infrastructure develops for society as a byproduct of the innovation forces unleashed during each new innovation wave. The railroad network was developed in the Age of Steel due to the invention of the Bessemer steel furnace. Electrical systems connected societies during the electrotechnology wave. Road and highway networks sprouted up across nations during the ascending phase of the automotive K-wave. Global telecommunications networks have appeared everywhere during the world's current innovation wave defined by computers, telecommunications and digital information – a trend that defines the modern world.

An extensive list of the previous western Kondratieff waves is as follows, which allows us to see that navies have played a crucial historical role in many of the previous prosperity booms due to the necessity to open markets or protect sea trade:

World Power	Date	Innovation/Leading Sector
Northern Sung	930	Woodblock printing
	990	National market formed
Southern Sung	1060	Public finance reform
	1120	Maritime trade expansion
Genoa	1190	Champaigne fairs
	1250	Black Sea trade
Venice	1300	Galley fleets
	1350	Pepper trade & Spanish gold
Portugal	1420	Guinea gold
	1492	Spices
Dutch Republic	1540	Baltic trade
	1580	Asian trade
Britain	1640	American plantations (rum, tobacco, slaves)
	1680	Amerasian trade (tea, coffee, sugar)
	1740	The Industrial Revolution, textiles, iron
	1790	Steam, railroads and industry
USA	1850	Steel, electric power, heavy engineering

1914	Automobiles, oil, mass production
1973	Information technology & telecommunications

From this list, one can see that Kondratieff waves have been characterized by revolutionary technologies or governmental, social, and marketing innovations that lowered costs or improved human productivity. They were basically the result of offering some type of greater efficiency to society that then impacted national economics:

- New transport modes that cut transportation costs and transport times (canals, railroads, roads, bridges)
- New trade routes and the products they made available
- Cheaper or faster communication-information flows (woodblock printing, telecommunications)
- More efficient, powerful and inexpensive sources of energy (steam, electricity, nuclear)
- Cheaper market clearing (sales channel) mechanisms (fairs, trade markets, reduced tax systems)
- Cheaper production techniques (steam motors, mass production)
- Any other "cheaper" improvement that added value

Kondratieff waves have been launched by opening up new sales territories, or by the discovery of new products that were ocean-sourced from foreign locations. They typically result from better productivities, such as economic efficiencies that lowered costs rather than increased trade. This harkens back to Reinert's observations on how countries traditionally became wealthy, which is by achieving cost efficiencies while increasing sales.

Sometimes Kondratieff waves constituted grand political or military innovations that brought about peace, which increases sales because peace provides safety for the flow of commerce (such as when the Roman Empire created a safe, stable economic zone absent of piracy by controlling the entire Mediterranean Sea), or resulted from government policy changes that lowered taxes and created larger economic efficiencies overall.

The extensive list of western Kondratieff waves clearly shows that national prosperity has at times been entirely dependent upon maritime trade flows (e.g., Black Sea trade, Galley fleets, Pepper trade and Spanish gold, Spices, Baltic trade, Asia trade, etc.), which explains why many of our top naval commanders spent a good portion of their careers defending sea trade routes and foreign ports. Navies have historically been used to secure the core of national prosperity and hegemony, where strategic port bases enabled them to protect trade lines and control choke points so that they could blockade the trade of competitors.

This list also makes it easier to understand why the U.S. used the naval forces of Commodore Matthew Perry (1853) to open up the Japanese market for its goods. In order to increase national sales and prosperity, U.S. gunboat diplomacy (1882) was also used to open up trade to Korea. U.S. naval forces and marines have also intervened in several Latin American countries – primarily in Central America and the Caribbean – to protect American business interests.

The United States is not the only nation to have used its naval forces to protect Kondratieff wave sources of income, or to open up channels for export sales and raw material purchases, which are typical Kondtraieff activities. In the first Opium War, Britain used its navy to bombard Chinese ports, leading to the Treaty of Nanking (1842) that opened five treaty ports (e.g., Hong Kong, Shanghai) for trade. After the Opium Wars, the U.S. used its naval presence to demand equal trade access too. In the Persian Gulf (19th–20th c.), the British imposed protectorates (Kuwait, Bahrain) to secure the oil trade while the U.S. Navy took over this function post-WWII. British gunboats enforced trade in Malaya, Burma, Singapore, while the U.S. took the Philippines in 1898. The British fleet also bombarded Alexandria, Egypt to secure the Suez Canal because it was vital for India trade. These interventions should viewed in light of a hegemon's desire to maximize its sales and sourcing channels as the Kondratieff cycles illustrate.

The world has progressed and prosperity – powered by innovation cycles – are no longer about trade flows and sales or imports of simple commodities such as spices, coal, textiles, or even oil. They are now mostly about technology or technological change and the benefits they bring to economies. The warning we need to consider is clear, which is that to become prosperous and stay prosperous a nation *must invest in developing new technologies or quickly adopting those found elsewhere.*

China, for instance, has learned this lesson well. China adopted best-in-class infrastructure and telecommunications when modernizing in the 1990s and 2000s in order to put it far ahead of its peers. Since that time, China has continued to invest heavily in innovation cycles and the early adoption of emerging technologies, making it a formidable competitor to the West after catching up and even surpassing the U.S. as the leader in several developing technologies, and special types of manufacturing.

Multiple reports and economic analyses from the early 2000s consistently showed that the U.S. was leading the world in information technology, semiconductors, biotechnology, aerospace and defense, and productivity growth. However, a groundbreaking 2023 report by the Australian Strategic Policy Institute (ASPI) found that China now leads in 37 out of 44 critical and emerging technologies, including:

- AI and Machine Learning
- Quantum Sensors and Communications
- Electric Batteries and Photovoltaics
- Hypersonics and Advanced Materials
- Biotechnology and Robotics

In many fields, China's top research institutions produce 5–9 times more high-impact papers than their U.S. counterparts. In some technologies, all top 10 global research institutions are based in China. China's dominance is attributed to long-term strategic planning, massive R&D investment, and talent acquisition – including researchers trained in Western countries. The U.S. has fallen behind, which translates into economic, financial, military and hegemonic risks.

Look at what has happened due to a national initiative by China, which is trying to secure its future prosperity. The U.S. now leads in only 7 out of 44 technologies, including vaccines, quantum computing, and space launch systems, compared to China's 37. *This should be a wake-up call to the United States.* Without prosperity, there can be no strong military.

As another wake-up call, due to American and European assistance in granting China favorable trade status and their firms relocating their manufacturing to China with technology transfers, China is now the manufacturer of the world. It has a lot of capacity it needs to keep running to keep its public employed, otherwise the Communist party will lose power. However, internal demand is presently collapsing due to China finally experiencing a long anticipated demographic decline (which will last for some years), and it is currently experiencing a real estate crisis with massive oversupply. China still has not fully developed a consumer-based economy (like the U.S.) to keep itself going, but is turning away from depending on the mercantile system that bases prosperity on exports.

China's situation is much like that of Japan in the 1990s when the Japanese manufacturing powerhouse hit a demographic wall and its real estate bubble burst. China has chosen the route of government spending and trying to export its way out of this crisis, which has massive deflationary implications for the U.S. and world. In other words, China is facing massive deflationary collapse (which you can see by its low bond yields), is trying to export its way out of crisis, and will be passing deflation onto the world.

Our brief study of innovation cycles reveals several critical lessons:

- Nations must invest in R&D and education to stay ahead of the curve.
- The early bird gets the biggest benefit from riding (adopting) an innovation wave.
- Countries that align policy with emerging technologies gain long-term advantage.
- Prosperity depends not just on technology, but on cultural and institutional adaptability.
- Down waves often follow peak success – requiring reinvention and resilience.

In the modern era, as long as a nation can obtain raw materials from the global market place that are necessary for its survival, technological leadership matters far more than resource control and commodity exports.

The Kondratieff, or innovation wave framework, demonstrates that national prosperity is not determined by geographic position, commodity resource endowments, or military conquest, but by the ability to identify, develop, adopt, and lead innovation cycles. Nations that now focus on fostering innovation or achieving early adoption of emerging technologies from others will prosper, while those that rely on resource extraction or cling to outdated theories (like the Mackinder geopolitical theory of geographic determinism) will fall behind.

The rise of China through strategic investment in emerging technologies, while the West has instead focused on financialization gains and geopolitical maneuverings, illustrates this principle perfectly. Jack Ma, founder of the Chinese firm Alibaba, during a 2017 interview said, "In the past 30 years, America has had 13 wars at a cost of $14 trillion. What if they had spent part of that money on building infrastructure, helping white-collar and blue-collar workers?" In other words, what did the U.S. get for that spending? Nothing – the American Neoconservatives who pushed for war have undermined the economy long-term with endless, fruitless wars and the country has deteriorated in various ways because the home front was neglected.

Ma said that the U.S. should focus less on military spending and more on domestic investment, especially in education and infrastructure. However, an even better focus should be on R&D and innovation where any consequential outcomes are manufactured inside the U.S. and then exported elsewhere. Many remember the Golden Age of American Innovation, which spanned roughly from the 1930s to the 1980s. It was marked by deep investment in fundamental research, corporate R&D labs and government-backed science – with such notable cases as Bell Labs, IBM Research, Xerox PARC, Hewlett-Packard, MIT Lincoln Laboratory, NASA, Los Alamos & Oakridge National Labs, General Electric Research Lab, Stanford research Institute – producing breakthroughs that shaped modern technology. How can this be re-initiated inside the U.S. in a new form that is appropriate to today's world?

The real economic competition now is not over controlling territory or seizing someone's resources, but over who will lead the next wave of technological innovation and reap the benefits that will come from it by leapfrogging their peers. This conclusion could be wrong, but evidence suggests that true national power and prosperity will come from this direction in the 21st century.

The Cost of Capital: Public Banking and National Prosperity

A second critical factor for national prosperity – beyond the large topics of industrialization and productivity gains, innovation cycles, increasing exports and markets, and so forth – is lowering the financial cost of business, namely, the interest rates within a nation charged to fund businesses and infrastructure projects. Debt will undermine everything a business has and must do, so it warrants that a nation lower the internal debt costs that fund its economy.

The financial and other costs of compliance with all sorts of *unnecessary regulations* is also a hindrance to prosperity that needs to be reduced, but here we are focusing on the financial cost of business funding and development.

This seemingly technical issue has profound geopolitical implications, as dominant powers have historically worked to suppress public banking systems in rival nations because they lowered the costs of capital in contender nations since this made them better competitors. Essentially, by maneuvering international interest rate policies to keep interest rates high in nations that competed with them, they tried to maintain their own economic advantages.

As Reinert demonstrated, nearly all rich countries used tariffs, subsidies, and state intervention to protect and nurture their fledgling industries during early stages of development. This list should include lower funding costs to subsidize businesses. Without such protection, established foreign competition – having already achieved lower costs of production through scale and experience – would kill off nascent industries in other nations before they had a chance to grow. "Free trade" was only adopted after a nation achieved industrial dominance, not during the catch-up phase.

Historically, many poor countries also stayed poor because Colonialism and free-trade policies (e.g., structural adjustment programs) forced those poor nations into commodity specialization, locking them into the low-growth traps that Reinert noted. The mainstream economics (like comparative advantage theory) that was preached to developing nations ignored the historical reality that industrial diversification, not raw materials, drives long-term wealth. In other words, industrialized nations tried to snooker poor countries into staying poor.

Confessions of an Economic Hitman, by John Perkins, further exposes how powerful corporations and U.S. government agencies in a parallel fashion manipulated developing countries for economic and geopolitical gain. Perkins describes how as an "economic hitman" he was tasked with convincing foreign leaders to accept massive infrastructure loans that would benefit American companies while indebting their nations. These debts, he claims, were used to exert control over countries' resources and policies, often leading to economic dependence and political instability. The mantra of "globalization" was actually a tool of empire rather than mutual development.

Similarly, preventing another nation from lowering its interest rates as much as possible to fund internal economic development at minimum cost is another type of containment strategy used by dominant world players.

We already know that Britain itself industrialized behind protectionist policies such as the Navigation Acts, but later attempted to prevent the United States and others from doing the same because it knew that industrialization was the path to wealth and it didn't want competitors. For example, Britain opposed U.S. tariffs in the 19th century to protect its industries because a commodity-exporting raw material supplying America would exist for their benefit whereas a manufacturing-centric United States would have become a formidable competitor. Britain therefore attempted to mislead the U.S. into adopting the wrong development strategy, and directly opposed U.S. moves to industrialize. After losing its former colony, it has consistently sought to humble and manipulate the U.SA. into a servant-like arrangement, some analysts even saying in order to ultimately recover it as an integral part of the British Empire.

A fundamental practice of geopolitics is to thwart another nation's trade routes, or inhibit/contain a rival's economic development in many ways so that your prosperity and predominance can continue for as long as possible, and use it to do your bidding. Britain applied this strategy not only through trade policy and by inhibiting manufacturing but also through systematic suppression of public banking systems in rival nations so that their cost of capital remained large, and thereby inhibited industrialization.

Anything that might help rival economies (such as Germany, Canada, and Australia) would weaken Britain's predominance, so using all sorts of misleading machinations Britain tried to send these nations in another direction, including raising their national interest rates and internal costs of capital. Britain has consistently acted this way throughout its history. Basically, it employed inhibition strategies to thwart the national banks of other nations, especially those within its sphere of influence. It had been consistently employing the Continental Theory of security that instructs nations to set other states up to fail, destabilize those that are rising, await for the right time to pounce and then ingest and absorb the fallen.

As Ellen Brown documents in *The Public Bank Solution: From Austerity to Prosperity,* Britain and later Anglo-American financial elites deliberately pressured or forced Germany, Canada, Australia, India, and many colonies to dismantle or privatize their highly effective public banking systems in order to keep the cost of capital high in those nations and preserve London's (and later Wall Street's) financial dominance. The public banks they hoped to remove offered these nations a significantly lower cost of development by providing credit at minimal or zero interest rates, keeping capital circulating domestically rather than flowing to foreign creditors who, as debt holders, might dictate to them what to do.

Germany's *Sparkassen* (public savings banks) and *Landesbanken* (state-owned regional banks) provided low-cost credit to small businesses and infrastructure projects, financing Germany's rapid industrialization in the 19th century. They kept capital circulating locally, reducing dependence on private banks and foreign creditors.

After World War I, the Treaty of Versailles (1919) forced Germany to dismantle significant portions of its sovereign financial system, deliberately undermining this efficient low-cost funding model to cripple the nation's economic recovery. U.S. and British reforms further privatized German banking, increasing the native cost of capital for business formation and operations.

The Bank of Canada (BoC), founded in 1935, was a public central bank that lent interest-free or at minimal interest to federal and provincial governments for infrastructure projects. It was responsible for funding Canada's post-WWII economic boom, including major projects like the Trans-Canada Highway. By avoiding private bond markets, the BoC kept national debt remarkably low and funded at lower interest rates. However, in 1974, Canada was pressured – reportedly through the Bilderberg Group and IMF influence – to stop direct lending to provincial governments, forcing reliance on private banks at market interest rates. These moves undermined the BoC with a catastrophic result: Canada's national debt exploded from $20 billion in 1974 to hundreds of billions of dollars today, with interest payments consuming an ever-larger share of government budgets.

The Commonwealth Bank of Australia (CBA) was another state-owned bank that issued debt-free money for public projects. It funded Australia's World War II effort without foreign loans and provided farmers and entrepreneurs with low-interest credit, dramatically reducing their cost of capital. Naturally, this accelerated economic development. However, the CBA was privatized in stages from 1991 to 1996 under pressure from global financial institutions. Today, Australia's private banks are among the most expensive in the world, with the cost of capital for businesses and infrastructure significantly higher than during the public banking era.

The pattern is consistent across nations and time periods. During the Colonial Era, the British Empire banned public banks in colonies. For example, India's Presidency Banks were privatized or restructured to serve British interests rather than Indian development. During the Post-War era, IMF and World Bank loans routinely required privatization of state banks as a condition of assistance. This occurred extensively in Latin America, Africa, and Asia, forcing nations into dependence on expensive private credit. In our modern era, global financial institutions continue to pressure nations to privatize public banking systems, arguing for "efficiency" and "market discipline" while ignoring the developmental advantages of low-cost public credit. The push towards privatization is all basically a plan to thwart low cost development.

The purpose behind these directives is simple. When nations control their own credit, they can fund modernization, full employment and infrastructure without austerity. That's why the model of state banking was systematically erased. If your cost of capital is less than others through low-interest public financing, and if you retire project debt after repayment, then over time your debt level does not grow. This puts you at a great comparative advantage versus other nations. State banks follow this policy, and China has also used this pattern as well for various infrastructure projects such as railways.

The economic logic is straightforward but profound. If your cost of capital is lower than your competitors' because your public banking system finances development with lower interest, you gain a cost and compounding advantage over time. Public banks can:

- Issue credit for productive projects at cost (covering only administrative expenses) and fund projects that private banks will not
- Retire debt after project completion and revenue generation begins
- Keep interest payments circulating within the national economy rather than flowing to foreign creditors (such as the British banking system)
- Stabilize the economy during crises by maintaining credit flows

This creates a virtuous cycle: lower capital costs keeps more money in the system circulating that can fund more investment, which generates more productivity and wealth, which strengthens a nation's competitive position. Conversely, nations forced to rely on expensive private credit, or *foreign credit*, face a vicious cycle of debt accumulation and slower development. The U.S. is the world's number one debtor nation, but is not crippled by its cost of capital because its benefits from its reserve currency status, which produces an inflow of foreign capital to fund its economic growth. However, it will definitely suffer economically if this is interrupted.

While this may initially seem tangential to naval strategy, as was the discussion on Kondratieff innovation waves, it is extremely relevant with profound implications. As we saw in our first chapter, naval strategies can be viewed by periods corresponding to which nation was the financial leader in the world. As financial power shifted from nation to nation, naval power and maritime strategies shifted as well to fit the economics. The nation that controls global finance not only becomes rich, but has a tendency to project power more effectively and can financially sustain longer conflicts.

Alarmingly, the United States is currently losing its position as the world's reserve currency. BRICS nations (Brazil, Russia, India, China, South Africa) are developing their own trade and currency solutions to replace the U.S. dollar in international transactions. Some nations have started trading gold and oil outside the USD system, and perhaps its position as the world's reserve currency is weakening in favor of gold. This represents a fundamental shift in global financial architecture comparable to the transition from British to American financial dominance in the early 20th century.

As this transition unfolds with the possibility of chaos during upcoming economic transformations, the question of public banking becomes strategically critical. Nations that can finance their own development through public banking systems will have an advantage over those using expensive private credit or foreign debt. China's state-owned banks, for instance, have smartly bypassed the U.S. dollar and financed its internal infrastructure development along with Belt and Road initiatives at costs that would be impossible through private banking systems.

To learn more, Ellen Brown's *Web of Debt* provides details on how private banks monopolized money creation, and explains how to create public banks that cut costs within states. Stephen Zarlenga's *The Lost Science of Money* examines historically suppressed public banking models and reveals frameworks that might be employed should a new financial system need to arise to deal with the debt and currency problems resulting from the U.S. losing its position of financial supremacy.

These models become particularly relevant as the current dollar-based system faces increasing strain, which will certainly affect the role of the military that has secured trade stability for all nations during the era of the Petrodollar. Nations seeking to escape dollar dependence and build resilient economies may need to rediscover and adapt these public banking principles, as should the U.S. simply to strengthen itself. The ability to finance national development at minimal cost, rather than enriching foreign creditors, could determine which nations escape international control and thrive in the emerging multipolar world order.

The U.S., under President Trump, is talking about replacing the fiat Federal Reserve dollar with a gold-backed Treasury dollar in order to accomplish some of these goals because the world is abandoning the dollar for trade and reserve status even though the United States is still the largest consumer-based economy that everyone needs to sell to. His move would flood the U.S. with foreign business relocating into the U.S. and help produce a lower cost of internal capital.

President Kennedy similarly ordered the Secretary of the Treasury to issue silver certificates (where the dollar could be converted into silver) but this was reversed after his death (on the 4th day of President Johnson being in office), and President Lincoln was killed after issuing "greenback" Treasury notes as legal tender, a move vehemently opposed by the British financial elites of his time who were collecting massive interest from U.S.

loans. While the USD remains the world's primary currency for trade, finance, and central bank reserves, cracks are showing in its dominance yet it will still be difficult to dislodge this premier status.

What we must recognize is that the connection of low-cost national debt to naval strategy is direct: a nation's ability to build, maintain, and operate a powerful military depends fundamentally on its economic strength and the cost at which it can finance military development and function. If a nation cannot sell its debt, it cannot finance its military. In fact, if a nation cannot sell its debt due to a loss of confidence in that government, then its finances will collapse.

Britain's naval dominance in the 19th century was built not just on shipbuilding prowess but on its ability to finance a global fleet through the City of London's financial system – while simultaneously denying rivals access to low-cost capital. However, it lost its naval status after WWII because of financial troubles.

Similarly, America's post-WWII naval dominance has been sustained by dollar hegemony, which allows the U.S. to finance military expenditures by printing the world's reserve currency. As this advantage erodes, nations with efficient public banking systems and low capital costs may find themselves better positioned to sustain naval development and operations.

China's rapid naval expansion, financed through state banks at minimal interest rates, demonstrates this principle in action. Meanwhile, Western nations burdened with high private-sector capital costs and massive public debts face increasing difficulty financing naval modernization.

This discussion on the cost of capital is not merely a technical financial issue – *money is a strategic weapon that dominant powers have historically used to suppress rivals and maintain their own advantage.* It flows to where it can find the greatest return. Just as we have seen many admirals try to thwart the trade of other nations, nations similarly act this way with money flows by creating situations that increase the development costs of competitors in order to maintain their financial dominance.

As the world transitions away from dollar hegemony, the nations that implement efficient public banking systems will be able to finance their development and military power at cheaper rates than others. The two will go hand in hand. The "lost science" of public banking may need to be rediscovered and adapted for any emerging multipolar world order, and particularly in the U.S. where the ability to finance national priorities at minimal cost could determine whether in the future it rises or declines.

The Mistaken Mackinder Theory Misguiding Policy

Before we introduce another type of financial warfare now aimed at the United States, we must clear away a 120-year-old intellectual obstacle that continues to distort Western strategy: Halford Mackinder's Heartland Theory.

Proposed in his 1904 paper to the Royal Geographical Society, "The Geographical Pivot of History," and later expanded in his 1919 book *Democratic Ideals and Reality*, Halford Mackinder argued that whoever controls the vast "Heartland" of Eastern Europe, Russia, and Central Asia would eventually command the "World-Island" (Europe, Asia, and Africa) and, with it, the planet. His famous dictum – "Who rules East Europe commands the Heartland; who rules the Heartland commands the World-Island; who rules the World-Island commands the world" – became the simple, seductive lens through which generations of British (and later American) strategists viewed Russia. He argued that controlling Eurasia's interior, namely Russia ("the Heartland"), would lead to a prosperous superstate that would attain global dominance, and eclipse Britain.

The theory was always fragile. It assumed:

- Geography is destiny,
- Land power would permanently eclipse sea power once railroads penetrated the Eurasian interior,
- and a single power could hold the Heartland together indefinitely.

History and technology have demolished every one of those assumptions. What has delegitimized Mackinder entirely are many facts such as the following:

- Air power, long-range precision strike, space weapons, and cyber-attack allow any serious power to hit the Heartland without ever setting foot in it.
- Nuclear weapons make large-scale land conquest of Russia politically and militarily impossible.

- Globalized supply chains, digital networks, and innovation cycles – not territorial control – now determine national wealth. Taiwan, South Korea, Japan, and Israel thrive with almost no natural resources; resource-rich empires (Mongol, Soviet) has historically fragmented and collapsed.
- Large land empires inevitably fracture under their own internal forces: the Mongol Empire split, the Soviet Union dissolved, and the same ethnic, linguistic, and regional pressures still operate inside the Russian Federation.

Three spectacular real-world tests have already falsified the theory:

1. The Mongol Empire (13th century) controlled the Heartland and then disintegrated.
2. Nazi Germany seized huge swathes of it in 1941–42 and was destroyed, and did not end up controlling the world.
3. The Soviet Union held the entire Heartland from 1945 to 1991 and still collapsed without firing a shot at the West. Neither did it end up controlling the world either.

At the time of its introduction, however, "The Geographical Pivot of History" had great appeal because it was the first theory to extend the scope of geopolitical analysis to encompass the entire globe. It was largely Eurocentric, which proved extremely attractive to the era's geostrategists who were worried about the future course of their empires. The theory basically told them that whoever controls Russia – with its vast resources of raw materials – will control the world.

This theory was especially attractive to the British Empire because its simplicity justified policies to intervene against Russia in all ways, giving policymakers, military leaders, and strategists something to keep busy with. Despite being discredited by logic and historical examples, the theory refuses to die – not because it is correct but because it is useful. It simply supplies a *simple, geographically deterministic justification for permanent confrontation with Russia*, which keeps many self-interested parties satisfied. It also is used as a justification for keeping Germany and Russia apart, thus preventing either from becoming a stronger competitor to Britain. It gives NATO a mission, defense contractors an enemy, and politicians a scary map to wave on television whenever they need an enemy.

As I pointed out in *The Art of Political Power*, national policy makers must strive to answer two questions: "How can I protect my nation, and how can I make it supreme?" They look for theories to provide insights to guide them, and part of Mackinder's appeal was that he was the only one at the time to offer a simple model for the entire world, and that model suggested conquering or containing Russia.

After Mackinder introduced his theory with its focus on the Heartland of Russia, it was a godsend to those who had spent their lives in the Great Game (1830s to 1907) that took Russia as an adversary. Now they had a reason to continue an antagonistic focus against their old adversary, which they believed was for dominion of the world, but now with the goal of conquest rather than simply containment. With Mackinder's theory as one's worldview, the vital question became, "How do we secure economic and political control of Russia?"

Since the Russian Empire had ruled most of the area from the Volga to Eastern Siberia for centuries, the question for those following the theory has essentially been how to seize its national resources from Russia. This, of course, is traveling in the exact opposite of Reinert's conclusions on how countries become rich, and totally ignores Kondratieff innovation waves that also build wealth while improving nations.

Yet, Mackinder's theory still helps guide British geopolitical moves even though it is easily discredited by history, technological developments that have decreased the importance of geography, and the fact that innovation is the key to prosperity rather than the capture of raw materials (which ultimately run out after a slope of ever-diminishing returns). For a national security apparatus looking for purposes, Mackinder's theory provides justification for active steps where officials are eager to be employed doing something. It justifies the use of weapons, specific diplomatic maneuverings, and complex machinations on a grand scale, which certainly appeals to the British.

The British evolved the basic idea from Mackinder that they must always keep Germany and Russia separated from friendship, which guided their actions in both World Wars. Furthermore, the British believe they must weaken both Russia and Germany economically while especially preventing Russia resources from powering Germany's industrialization. If relations grow warmer between Germany and Russia, trade barriers will be lifted, regulations eased, travel and tourism will increase, transactions between each other can be made in their own currencies, and there will be no need for U.S. military bases or expensive U.S. weapons. Thus, there will be no need for NATO should Russia and Germany become more friendly with one another.

Overall, the British want to keep these two nations apart and keep Russian influence (good relationships) out of Central Europe in general. Therefore, British strategy ever since has been fixated on preventing a linkup between Russia and Germany and thwarting both nations' development. Together with the U.S., these two nations have promoted the climate hoax within Germany to weaken its economy by damaging its energy supply and industrial base.

As for the Ukraine-Russia war, border conflicts have occurred within Europe for centuries, including the recent breakup of Yugoslavia, the 1993 division of Czechoslovakia, the 2006 establishment of Serbia and Montenegro, and the dissolution of the Soviet Union in 1991. It has been common throughout European history for the great powers to negotiate spheres of influence and settle conflicts through territorial compromise such as border adjustments or the creation of new states to avoid a larger, more destructive war. Therefore, the current EU clamor about the Ukrainian-Russian conflict, which was certainly provoked by the West, reveals deeper hidden motives other than simply the border dispute that it actually is. It is really an economic war by Western powers aimed at destroying Russia and then dismantling it.

In invading Ukraine to prevent Russians from extermination, Putin has actually shown restraint in waging a "special operation" rather than a U.S.-style "shock-and-awe" destroy-everything campaign at the cost of Iraqi-like civilian casualties. Putin seeks specific goals like ending the genocide of Russians inside Ukraine, clearing out bioweapons labs, and preventing Ukraine from NATO membership. Russia is certainly not attempting to achieve a fundamental restructuring of the global order. There are no imperial ambitions otherwise it would have set out to conquer Ukraine in weeks. It certainly isn't fighting for territory because if it was then the Russians would have bombed power plants, water supplies, communication hubs, railroads and infrastructure on day one to so that Ukraine would collapse. Then they could take it over in an extremely short time, but instead they have proceeded more gingerly.

China certainly understands the strategic implications, which is why it will not let Russia lose any war involving NATO, knowing it would be next. Chinese Foreign Minister Wang Yi told EU foreign policy chief Kaja Kallas in July 2025 that China does not want to see Russia lose the war in Ukraine because a Russian defeat would allow the United States to shift its full strategic focus toward China. In essence, "We are not prepared to let Russia lose. We know we would be next." Kaja Kallas declared in September 2025 a new proposal aimed "to speed up the phase-out of Russian liquefied natural gas (to be complete) by 1 Jan 2027," which would raise natural gas costs to Europe, harming EU nations greatly simply in order to cut Russian revenues and its ability to raise funds for the Ukrainian conflict. The rationale is, "Cash from energy sales keeps Russia's war going. We won't stop putting pressure on Russia until it ends its war." Bureaucrats are ready to inflict upon unwilling populations a new round of destruction based on an imaginary enemy and hatred for the Russian Communists of the Stalin-Khrushchev generation who no longer exist. The Russians themselves are supposed to suffer again for having previously suffered under Communism, which they did not want.

Behind the Western narratives to EU populations, Western analysts (as stated) actually hope to conquer Russia and afterwards use its fragments to encircle and then conquer China. China, of course, knows this.

The idea of connecting geographic data with national policy is so simple to understand that whether right or wrong, few can resist the attraction to Mackinder's geographical simplicity and how it impels them to act. Similar geopolitical theories based on geography have sometimes gained sway in geopolitical circles, but geography is just one of many factors affecting supremacy and not necessarily the primary concern.

By focusing on Mackinder, the United States and EU have a serious problem with strategy and institutional inertia. The fact that both the EU and United States are facing their own financial problems of potential debt defaults and economic collapse from poor internal political and geopolitical moves (e.g., excessive social spending, unrestricted immigration that doesn't assimilate, ridiculous climate change initiatives, COVID missteps, etc.) is testament to this fact, which are problems that in part originate from shipping their manufacturing bases offshore and industrializing China so that it has become a major competitor. Western powers are clamoring about Russia as a threat, yet did everything possible to make China the formidable rival.

Today the Heartland obsession has definitely become actively counterproductive. While Western capitals fixate on containing a Russia that no longer seeks global ideological domination, China is quietly winning the real 21st-century game: leading the next innovation wave, building critical supply chains, and financing its rise with sovereign, near-zero-interest credit through state-owned banks.

Nevertheless, Mackinder's ghost still haunts Western war colleges and foreign ministries, but it is leading them in exactly the wrong direction. Geography informs strategy; it no longer dictates fate. Prosperity and power now flow to the nation that masters technology, manufacturing, and low-cost public finance – not to the one that plants a flag in the middle of the Eurasian steppe.

Nonetheless, despite its flaws people still cling to Mackinder's theory because it supplies reasons to attack Russia, whose $75 trillion in assets are currently seen as a solution to the insolvency of European national debts and the financial disaster these nations have produced through missteps. EU nations also hope to seize $300 billion in Russian bank deposits, 80% of which belongs to private Russian citizens rather than being Russian state assets. The motives for demonizing Russia are, in actuality, primarily financial.

Western government debts are unpayable, EU nation states are effectively bankrupt or facing bankruptcy because their economies are collapsing, the world is de-dollarizing, and the West's monopoly over world financial institutions is slipping. All the military power amassed or land conquered in hopes of absorbing resources will not change this fundamental reality.

Some critics who rightly point out that Mackinderian analysis should be ignored in light of such truth are drowned out because far too much literature repeatedly trumpets the importance of preventing Russian dominance – a legacy of Russia once being an expansionist Communist country. Because of this, decades of strategy based on mistaken Mackinder cannot be easily abandoned even though Russia is now just another European nation. Just as once-enemies Germany and Japan have become friendly allies, Russia is also striving to normalize and integrate with the

Western world, but has this intellectual bias against it. But without an enemy, what are defense contractors, weapons suppliers, and geostrategists to do?

Nicholas Spykman's Rimland Theory (1940s) directly challenged Mackinder by arguing that the coastal regions surrounding the Heartland – the "Rimland" (Western Europe, Middle East, South Asia, East Asia) – were actually more important than the Heartland itself because the Rimland contained most of the world's population, industrial capacity, and economic activity, making it more valuable than the resource-rich but underdeveloped Russia. The Rimland also had access to both land and sea power. This suggests that a would-be hegemon should try to gain control over South East Asia and these other areas instead of Russia! Think China.

As previously explained, raw material resources are necessary for a nation's survival (nations fight wars to gain them) but where supply chains are secure the real competition for global leadership today is taking place in fields such as AI, advanced computing, communications, specialized supply chains, biotechnology, genetic engineering, next generation energy systems, and creating infrastructure for the next innovation wave.

South Korea, Japan and Taiwan, for instance, became wealthy despite their tiny size and minimal natural resources by focusing on manufacturing and innovation whereas the Soviet Union had vast resources but remained economically stagnant. How can it be considered the supreme power Mackinder warned against? Silicon Valley's innovation has created more wealth than all of Russia's minerals, too! The key to prosperity is not control of Russia's resources and raw materials but innovation, technology and manufacturing.

Alarmingly, the West's financialization and de-industrialization policies have led to its own decline, strip-mining the middle class, while the Soviet Union collapsed due to its own self-inflicted political-economic problems despite controlling the Heartland. China, however, is rising through its Belt and Road Initiative (Rimland strategy) together with innovation and manufacturing, not through conquest of Russia but through its assistance and friendship. China has understood the real game and is setting out to strategically win it.

Future geopolitical supremacy will not be about seizing central Asian resources – raw material control – or military confrontation to control Russian territory and confer dominance. China has understood the real game – leading innovation cycles and creating critical product supply chains – and is progressing along these lines while the U.S. stays passive in these necessary steps.

The path to prosperity is not through conquering Russia or by controlling geography, but through fostering innovation, investing in R&D and education, maintaining a strong industrial base, and positioning to ride the next wave of technological revolution. This follows from Reinert's observations and the lessons of the Kondratieff cycle, which makes Mackinder's geographic determinism a dangerous distraction from the real sources of national power and prosperity in the 21st century.

In summary, Mackinder's Theory has been discredited because of countless flaws. Yet it persists because it provides convenient justification for anti-Russian policies driven by various institutional interests, financial desperation (European debt crisis), and the psychological need for an enemy as well as the need to do something.

This is a dangerous delusion that ignores the real sources of national power in the modern world, yet certain British and American interest cling to it as justification for predetermined objectives. It justifies military budgets, provides a purpose for defense establishments (without Russia as a threat NATO lacks a mission), its geographical determinism is easier for policymakers, and provides a rationalization for predetermined objectives, such as the British desire to keep Germany and Russia separated from one another. The theory seems to persist not because it's correct, but because it justifies what powerful interests already want to do. Yet a war with Russia *will not achieve* any of the desired economic objectives.

Again, because this is so important to understand, the more important focus than capturing resources from Russia is the factor of R&D innovation, and fostering the development of innovation cycles. The country that can develop a leading innovation wave or adopt it at the earliest obtains the lead in prosperity, while resources and raw materials are simply a necessary fuel but not commodities that bring a lasting wealth of strategic riches.

While the Mackinder theory and its geopolitical imperative has been thoroughly disproved, its implication that "we must marginalize Russia, contain Russia, control Russia, defeat Russia, conquer Russia, divest Russia of its wealth" is just one more alluring slogan pushing everyone in the wrong direction. Russia is no longer Communist nor an enemy of the West, at least no more than Japan or Germany.

Russia is actually less of a threat because it is hobbled when it comes to industrial development, but the one nation that has taken off like a rocket in this respect is China, mostly due to U.S. policy errors. China's recent ascension challenges Mackinder directly because Beijing focuses its development efforts on coastal Asia (Mackinder's "Rimland") via its Belt and Road Initiative, using sea routes and land-based economic corridors rather than Heartland conquest for supremacy.

China is expanding into Central Asia (the Heartland) while also building its naval power, which challenges Mackinder's simplistic land-sea dichotomy. China's "String of Pearls" strategy (Indian Ocean bases) ensures maritime leverage alongside land dominance.

The Thucydides Trap: Historical Patterns of Great Power Rivalry

Mackinder's theory is just one of the many reasons that there is a great antipathy towards Russia. A recurring pattern in history provides us with another reason: the growth of one power inevitably causes fear in the established hegemon, making conflict nearly unavoidable. This explains much of the historical and contemporary Western antagonism toward Russia.

In his *History of the Peloponnesian War*, Thucydides (c. 460-395 BC) identified what he believed to be the fundamental cause of the devastating 27-year war between Sparta and Athens: "What made war inevitable was the growth of Athenian power and the fear which this caused in Sparta." The fundamental driver, he argued, was Sparta's deep-seated fear of Athens' rising power and imperial ambitions – a fear mixed with envy and distrust that made a major conflict inevitable.

This pattern provides yet another valuable lens for understanding contemporary geopolitics. Russia's size, military strength, and resource wealth alone are sufficient to make it a target of the current world leaders, regardless of its ideology or actual nature of being or not being a threat. This mirrors how Britain viewed Russia with antagonism in the 19th century – not because of any specific provocation, but simply because Russia was big and therefore an affront to British imperial dominance.

Russia is large, militarily powerful, rich in resources, and economically recovering from the constraints imposed by decades of Communism. In short, it is big and powerful, which is reason enough for the current world hegemon and other contending leading powers to oppose it. China also fits this profile, but Western powers have chosen to focus on Russia. No doubt China is next.

Nations commonly promote narratives as to why competitor nations should be considered enemies. Basically, the existence of a strong Russia is antithetical to America's strategic objective of full-spectrum dominance and geopolitical primacy. Russia – because of its size, resources and military power – stands in the way of unrivaled U.S. preeminence everywhere. For that reason alone, the U.S. wants to conquer Russia whereas for the EU, conquering Russia is seen as a solution to giving those nations a financial life line. The strategic calculus also put forth is that if the U.S. neutralizes Russia first, it can then use the resulting geopolitical fragments to next encircle and breakup China, which is another hegemony contender.

During the Cold War, the West opposed Russia because it was Communist. Now Russia is simply a growing rival that cherishes its policy independence and refuses to bow to Western hegemony. The previous narrative was that Russia was a godless, Communist country, which made it an automatic enemy. Today it is an Orthodox, non-Communist nation, yet Western hostility persists – even though the West no longer harbors animosity toward Germany or Japan, both of which waged wars that killed millions.

As we've previously seen, there was a British precedent to all this antipathy. The prominent American economist and public policy expert Jeffrey Sachs gave an interview to journalist Tucker Carlson on this. In that interview, he talked about the historical animosity towards Russia as lacking any real foundations. Sachs explained that in 1840, before the Bolshevik Revolution, the world hegemon was Britain – and British elites hated Russia then too.

Sachs cited historian John Howes Gleason, who in 1950 investigated why the British harbored an animosity toward Russia. In 1853, Britain went to war with Russia in the Crimean War (1853-1856), a major conflict fought primarily on the Crimean Peninsula between the Russian Empire and an alliance of Britain, France, the Ottoman Empire, and later Sardinia. The

British and Russians had been allies in defeating Napoleon, yet just 25 years later they were enemies. Why?

Gleason went through all the speeches by British leaders, House of Commons debates, and articles in intellectual magazines, searching for a rationale. What had happened? *He found no substantive reason for British hatred of Russia in the mid-19th century.* There was no reason for the change. Russia was simply big, and therefore an affront to the British Empire simply because of its size. This alone drew Britain's derision and the desire to clip Russia's wings. In their quest for global dominion they saw Russia the chief rival.

This rivalry resulted in "The Great Game," a term popularized by British officer Arthur Conolly and romanticized by writers like Rudyard Kipling in *Kim*. It referred to the 19th-century geopolitical competition between the British Empire and the Russian Empire over influence in Central Asia, particularly Afghanistan and surrounding regions. Britain feared that Russia's southward expansion toward India – the crown jewel of the British Empire – would threaten its colonial dominance. However, the Great Game was not just about India but had been going on for centuries, well before 1840. Russia indeed sought warm-water ports and influence over Central Asian khanates, pushing toward the borders of British India, but had no intention of invading any part of India. *This notion was an invention of the British mind* by bureaucrats who needed something to work on.

Princess Olga Paley of Russia, widow of Grand Duke Paul Alexandrovich of Russia (the youngest brother of Tsar Alexander III and the uncle of Tsar Nicholas II), noted in her autobiography, *Memories of Russia 1916-1919*, that it is strange that Russians have sometimes been fooled into trusting the British for "in the history of Russia, the animosity of England traces a red line across three centuries." Britain is always concocting machinations against it as a form of great power rivalry, just as it is always trying to covertly subjugate the U.S. once again. In *How the British Invented Communism (and Blamed It on the Jews)*, there is even substantial credible evidence that the British secret services played leading roles in the French and Russian Revolutions, and then performed an extensive rewriting of history as a coverup for its involvement, including its role in the overthrow of Czar Nicholas and rise of Communism that it fostered because Marxism served the Empire.

The British fabricated the absurd notion that the Czar would invade India through the Khyber Pass – that Russia wanted to march across

Afghanistan into India to fight Britain. This idea was absurd, nevertheless this Russophobic idea made the British view Russia as the greatest threat to their empire.

By 1853, Lord Palmerston (Henry John Temple, 3rd Viscount Palmerston), who dominated British foreign policy from 1830 to 1865 when Britain stood at the height of its imperial power, engineered a pretext to go to war with Russia, which became the Crimean War. In 1844, Tsar Nicholas I had visited London for private talks with British officials, including Lord Aberdeen, reaching a secret verbal agreement – the Nesselrode Memorandum – to cooperate in maintaining the status quo in the Near East. Russia had told the British, "we don't want to fight you," but the British pushed ahead in creating an excuse for the coming conflict. Even in the early 1850s, Nicholas I continued to express that Russia had no hostile intentions toward Britain and hoped to resolve disputes diplomatically. However, because of the underlying British mindset any Russian actions in the Balkans and Black Sea were interpreted as expansionist and threatening, interpretations that ultimately lead to war. Everything Russia did was interpreted as aggression by Britian. Miscommunication, mistrust, and shifting alliances led Britain and France to declare war on Russia in March 1854.

Russia lost the Crimean War. Under the Treaty of Paris (1856), Russia ceded territories and was forbidden from maintaining a navy or military fortifications in the Black Sea, severely limiting its strategic influence in the region, which is exactly what Britain had wanted. Russia's defeat weakened its image as a dominant European power, exposing its technological and logistical backwardness compared to Britain and France. Lord Palmerston had engineered this conflict to banish Russia from the Black Sea (its fortifications and fleet) and reduce it to a third-rate power, and succeeded.

Today, history appears to be repeating itself. Because of the Ukraine War and the need to clear their unpayable sovereign debts, European nations are hysterically claiming that Russia is planning to attack EU countries, which is also absurd because it would gain nothing, and are drumbeating that war with Russia is inevitable when they are the ones trying to cause it. In fact, Russia's Foreign Minister Lavrov has stated that Russia has "never had and does not have any such intentions" to attack NATO or EU countries, though "any aggression against my country will be met with a decisive response." Russia has nothing to gain from conquering countries in Europe.

Russia has no intention of conquering Europe and did not invade to take Ukraine either, but did so to take the Donbas. In my view the Donbas, comprised largely of a Russian speaking population that was being exterminated by a Nazified Ukrainian military, should be allowed to separate from Ukraine to produce a lasting peace, especially since it had already voted in free elections to join with Russia. *This separation would parallel the same final solution for the breakup of Czechoslovakia into the Czech Republic and Slovakia, as well as the breakup of Yugoslavia along ethnic lines. In other words, this has been done before.*

The Ukraine conflict is not about Ukraine – it is about Russia itself. It was provoked by the West and is about the desire to conquer Russia and fragment it to solve a host of Western financial difficulties and geopolitical objectives. It is a war backed by Neocons and NATO where Ukraine is being used as a sacrificial lamb to weaken their Russian archnemesis that is no longer Communist and expansionist, so no longer an opponent. The EU support of Ukraine has never been about Ukraine, but about the conquest of Russia. The Western powers do not actually care about Ukrainians being killed because they are expendable pawns on their geopolitical chessboard. However, history has several times shown the West interfering with Russia (e.g. Hitler and Napoleon), with campaigns often remembered for their ultimate failure.

The problem for the West is that war drums make it impossible for nations to sell their sovereign debt – which is exactly their key financial dilemma at present – because interest rates and inflation rise on such fears. Western powers are already struggling to sell their debt and are making it worse by talk about war, which would enable them to steal Russian bank funds and declare their debts null and void as they have done in the past.

Currently, Europe spreads propaganda that Russia is weak, but the military and economic reality is the opposite. Just as Russia tried to avoid war with Britain before the Crimean War, its actions in Ukraine are being interpreted as expansionist and threatening – an interpretation that may lead to conflict. It is actually the Europeans who seem to be warmongering with such views. Lord Palmerston maneuvered to banish Russia from the Black Sea and turn it into a 3rd rate power, and Britain is hoping that a Russian conflict today can similarly cripple Russia. Yet each time moves are made to weaken Russia through sanctions and other measures, it comes out stronger and the EU weaker.

Interestingly, because a unified German Empire was arising under Otto von Bismarck after the Crimean War, France sought an alliance with Russia in the early 1890s because the Germans were gaining in power. Think about this: following the diplomatic fallout of the Crimean War in viewing Russia as an enemy, the French began seeking an alliance with Russia against Germany in the early 1890s! This is one of thousands of instances showing that enemies can become friends and friends can become enemies in geopolitics; there are no perpetual allies nor perpetual enemies – nations follow their own interests to ensure their survival and prosperity. Also, the reasons for hatred or war can arise from misinterpretation, underlying prejudice, or deliberate fabrications such as false flag events.

A modern historical error of misinterpretation was seen when Robert McNamara's announced profound regret over the Vietnam War where 58,000 Americans were killed. Before his death, he acknowledged that the U.S. had misunderstood the nature of the conflict, mistakenly viewing it as a Cold War battleground from assuming that North Vietnam was merely a puppet of Soviet or Chinese Communism. In reality, he came to believe it was primarily a *civil war* with deep historical roots.

Britain's Prime Minister Tony Blair also apologized for Britain's part in the Iraq War against Saddam Hussein, claiming, "I can say that I apologize for the fact that the intelligence (on weapons of mass destruction) we received was wrong … I also apologize for some of the mistakes in (post-war) planning and, certainly, our mistake in our understanding of what would happen once you removed the regime," which was to subject the people to sectarian violence and instability. Leaders misinterpret geopolitics repeatedly, and rally their nation to the cost of millions of lives lost needlessly due to their mistakes.

Jeffrey Sachs relates that the Deep State, CIA, military contractors, and neoconservatives have maintained hostility toward Russia since 1945, even though the United States allied with the Soviet Union to defeat Hitler. After World War II, the U.S. began arming itself against Russia almost immediately, initiating the Cold War against Communism. In December 1991, when the Soviet Union finally ended, Sachs was present at a meeting where Boris Yeltsin indicated that Russia no longer wished to follow Communism and said, "We want a normal country, a normal economy."

When the Soviet Union collapsed in 1991 ending the Cold War, the West – particularly the Deep State, CIA, and military-industrial complex – immediately shifted from containment to dismemberment, hoping to break Russia into three different parts. Rather than accepting Russia as a "normal country" (as Boris Yeltsin requested), the U.S. pursued NATO expansion eastward, breaking explicit promises. This expansion, formalized in 1999 with Poland, Hungary, and the Czech Republic joining NATO, was designed to surround and fragment Russia. The idea became that the United States should be the only major power on the planet. China was not even a consideration at that time! The focus remained solely on countering Russia.

By 1992, the Deep State, CIA, intelligence agencies, Pentagon, armed services committees, and military contractors, etc. embraced the concept of a single world hegemon – the unchallenged primacy of the United States. This was variously called neoconservatism, U.S. unipolarity, primacy, full-spectrum dominance, or hegemony. Naturally, these organizations required Russia as an enemy to justify their budgets, expenses and policies. The rule is, without an enemy one must still be created to justify our organizations. In time, the military focus is sure to switch to the ascension of China.

As Sachs relates, When President Clinton took office, various nations approached him claiming they needed protection from Russia. He asked, "What for?" – the Cold War was over and the Soviet Union destroyed. But not for the Deep State, which still needed an enemy and still wanted to eliminate Russia in 1993.

The Deep State used Bill Clinton to usher in NATO expansion. In 1994, Clinton endorsed the eastern expansion of NATO through the Partnership for Peace, with formal expansion occurring in 1999 when Poland, Hungary, and the Czech Republic joined the alliance. The expansion of both NATO and the European Union to Russia's detriment, threatening it in various ways, has been the West's policy posture for decades.

The record is clear that the U.S. promised *not* to move NATO eastward, yet Clinton initiated the enlargement that threatened Russia's security and violated those promises. In his 1997 book, *The Grand Chessboard*, Zbigniew Brzezinski, a highly influential diplomat and geostrategist, cautioned that "the most dangerous scenario would be a grand coalition of China, Russia and perhaps Iran, an 'anti-hegemonic' coalition, united not by ideology but by complementary grievances." So in 1997, Zbigniew Brzezinski warned that NATO expansion would provoke Russia into forming an "anti-hegemonic coalition" with China and Iran.

This is exactly what happened due to stupid moves and bad strategy on the part of western powers, particularly the U.S. and Britain. *Their aggressive posture pushed Russia into the arms of its strategic rivals, creating the very threat U.S. policies claimed to prevent.* As it now stands, Russia, China, Iran, North Korea and Pakistan are in a powerful alliance, and the BRICS have formed as well.

In his book, Brzezinski also wrote, "Ukraine, a new and important space on the Eurasian chessboard, is a geopolitical pivot because its very existence as an independent country helps to transform Russia. Without Ukraine, Russia ceases to be a Eurasian empire." He argued that Ukraine's independence was crucial to limiting Russia's imperial ambitions, that is Russia regained control over Ukraine it could reassert dominance over Eurasia, and therefore U.S. strategy should support Ukraine's sovereignty (whereas Ukraine had never been an independent nation throughout history) to prevent a resurgence of Russian imperial power. The U.S. could also use Ukraine to buy time in keeping Russia the bad guy so that America could still maintain a large influence over Europe. This explains some of the voices over the conflict.

Because the Cold War is over and Russia is no longer an expansionist Communist power, Chinese colonels Qiao Liang and Wang Xiangsui observed in *Unrestricted Warfare*: "The American soldiers who had lost their opponent due to the collapse of the former Soviet Union are vehemently searching for a reason not to allow themselves to be 'unemployed.' ... It should not be said that as soon as there were no longer two armies facing off against each other that American military circles and the American Congress would produce an empty feeling at having lost their goal. The result was that without an enemy, one still had to be created."[2]

Jeffrey Sachs emphasizes that all this history shows that we are not witnessing short-term presidential decisions but a long-term strategy directed by politicians, bankers, the intelligence community, and the military establishment. The current moves against Russia are not reactions to current politics but a misguided project dating back more than 30 years – one that will likely backfire, *just as we have seen other mistakes along the lines of*

---

[2] *Unrestricted Warfare*, Qiao Liang and Wang Xiangsui, (Echo Point Books & Media, Vermont, 1999), p. 106.

grand strategy throughout history. Central bank advisor Martin Armstrong was arrested in 1999 in response to refusing to join one such geostrategic plan to financially topple Russia, which failed miserably, and his story follows later.

The pattern from Lord Palmerston's strategy to surround Russia in the Black Sea region, to the breakup of the Soviet Union, to the U.S. strategy of encircling and dismantling Russia, reveals a consistent thread: the fear and resentment of a rising or resurgent power. This is the recurring historical pattern identified by Thucydides that the growth of one power inevitably causes a reflexive reaction by the established hegemon that makes conflict nearly inevitable. This pattern explains both historical and contemporary Western antagonism toward Russia – at least a large portion of it.

Essentially, Russia's size, military strength, and resource wealth alone are sufficient to make it a target of the current hegemon (the United States), regardless of ideology or actual threat. This mirrors how Britain viewed Russia in the 19th century – not because of any specific provocation, but simply because Russia was big and therefore an affront to British imperial dominance. Western hostility toward Russia has persisted across radically different political systems. During the Cold War, the justification was anti-Communism. Today, despite Russia being an Orthodox Christian state and non-Communist, the hostility continues – even though former enemies Germany and Japan have been rehabilitated. This suggests ideology is merely a pretext, not the true cause.

An additional factor, Western military and intelligence institutions also require Russia as an enemy to justify their budgets and existence. As Chinese military strategists observed, American military circles needed to create an enemy after the Soviet collapse to avoid "unemployment." This institutional imperative helps drive policy more than any genuine security threat.

In short, the pattern from Lord Palmerston's strategy to surround Russia in the Black Sea region, to the breakup of the Soviet Union, to the U.S. strategy of encircling and dismantling Russia, reveals a consistent thread: the fear and resentment of a rising or resurgent power, and the need to have an enemy. As Thucydides observed 2,500 years ago, the fear of another power ascending makes conflict seem inevitable – *unless leaders consciously choose a different path.*

The current Ukraine conflict mirrors the Crimean War: European nations manufacture hysteria about Russian invasion plans (just as Britain did in the 1850s), Russia denies any hostile intentions for Europe (just as Tsar Nicholas I did), yet the West interprets Russian actions as expansionist and threatening. What complicates the antagonism towards Russia this time is driven by organizational needs for continued military spending, geopolitical objectives on the part of the West, and impending European financial bankruptcy (from its welfare state, Green agenda and immigration policies) where the solution is war as an excuse to default upon unpayable public debts that have spiraled out of control, and install capital controls and a new currency. War is an easy excuse for deficits, censorship, and redirecting public anger from domestic elites towards a foreign adversary. Can we actually choose a different path other than a European war?

Western antagonism toward Russia is not a response to specific Russian ideology or actions, such as its military operations in Ukraine, but rather a structural feature of hegemonic competition, a response to the need for purpose and search for a solution to national debts. The pattern spans 180 years, from British imperial rivalry through the Cold War to contemporary NATO expansion, driven by fear of a large, powerful rival and institutional interests that require a permanent enemy to justify their existence. This long-term strategic mindset, embedded in Western political and military establishments, risks repeating catastrophic historical mistakes.

Martin Armstrong On Western-Russian Antagonism

Global financial manager, capital flow expert, central bank advisor and geopolitical analyst Martin Armstrong has provided a yet more compelling explanation for the current antagonism toward Russia. In his extensively documented work, *The Plot to Seize Russia – The Untold Story*, Armstrong reveals a covert geopolitical strategy from the 1990s and early 2000s aimed at seizing control of Russia following the collapse of the Soviet Union.

Armstrong explains how New York bankers invited him to invest $10 billion into Hermitage Capital Management to fund Russian regime change (oligarch Berezovsky to replace Boris Yeltsin) and target Russian assets during the chaotic post-Soviet transition. The fund's strategy was intended to pave the way for Western-backed leadership in Russia, and then all of its resources would trade through the New York dealing desks where

Armstrong would likely make a $100 billon return on his $10 billion. Armstrong was offered a letter of guarantee from Republic National Bank's holding company, but refused to participate, seeing the plan – which entailed Russian regime change – as morally and geopolitically dangerous. His economic models showed that Russia would collapse and this plan would fail.

He later interpreted the proposal as part of a broader regime change strategy involving Western intelligence agencies, Russian oligarchs like Boris Berezovsky, and elements within the Clinton administration, which he detailed in his book. Armstrong ties this plot to the rise of Vladimir Putin, who was allegedly chosen by Boris Yeltsin to resist such foreign interference, protect Russia, and put an end to the political meddling by the Russian oligarchs. Armstrong also states that these secretive financial moves helped shape the modern geopolitical divide between Russia and the West, with long-term consequences for global stability.

Having failed to divide Russia into several territories and strip it of assets, Western powers still aim to weaken and fragment it, and then use those pieces against China, the next target, and China understands this objective. These two countries, Russia and China, have a complex relationship of co-dependency but the Western pressure and meddling has inadvertently pushed them closer together, which is the exact opposite of what geopolitical strategists intended. Richard Nixon and Henry Kissinger opened the doors to China in order to split it apart from Russia, and now we've glued them back together again. China recognizes that if Russia were conquered, it would become the next target of a similar destabilization-division campaign so it is, and will be, supporting its neighbor.

Armstrong explains that European governments have become bankrupt from years of socialist policies, unchecked immigration, useless climate initiatives that damaged EU economies, COVID missteps and negative interest rate policies – all failed strategies. These governments cannot allow their populations to discover that government interest payments, public debts, and social welfare costs are out of control and unsustainable on their watch while their economies tumble. Endless debt cannot restore their welfare systems, but the public will discover this truth and Eurozone instability will spread like a wildfire across Europe.

Government debt levels across individual European countries have risen to levels where mathematically they cannot be repaid while interest payments are crowding out social welfare costs. Other nations don't want to buy any new sovereign debt that EU nations issue, which guarantees defaults and then revolutionary tendencies within those nations. Pension funds and insurance companies are also in crisis after years of negative interest rates (2014-2022) because EU law compelled them to hold government bonds, which guaranteed losses. The EU pension funds on average must have 70% government debt. Hence, the EU pension funds and banking systems are undermined. Defaults are mathematically inevitable, which will produce economic chaos and trigger civil unrest in time. Economies are also on the decline, so war is seen as a way out to distract the populace and apportion blame to someone else.

No one wants to be blamed for the crisis that is coming. No central banker, politician or bureaucrat wants to lose their power and pension. Governments will topple if their politicians and central bankers are finally forced to assume blame for the collapse of their financial systems and broken economies. Economies are heading towards a deep recession or even depression, so the EU needs an external enemy desperately. They need to create a crisis to blame for the coming financial collapse and economic deterioration, which are happening of their own weight.

The EU nations need to create a distraction or diversion so that the attention of their voters is pulled away from sharply deteriorating socioeconomic problems the leadership has created, and so they are shouting, "Russia! Russia! Russia! Russia may invade us!" At the same time, they have turned authoritarian and now censor free speech, trying to clamp down on any flow of information that disagrees with government viewpoints as well as criticizes the government or its policies. The historical principle is that a government in trouble turns authoritarian and then makes moves to cancel civil rights, free speech, and freeze capital movements (by instituting capital controls) if necessary.

Panic might even cause the EU nations to spend a massive amount of money and create even more debt but eventually it will all default when no one wants to buy the new debt and the EU currency falls in collapse, so the EU must introduce a new digital currency (and perhaps universal basic income) and reset the financial system in a way that still keeps the elites in control. It will not be too much for the EU to go as far as rigging elections and starting wars so that the elites don't lose power but can deflect blame elsewhere for their missteps. However, if a nation goes to war then no one will buy its debt because interest rates will rise, which will immediately lower the value of that debt.

According to Armstrong and his discussion with high officials, the EU needs a war in order to cancel national debts along with the Euro currency just as European nations have done repeatedly throughout history. Essentially, the EU is economically deteriorating and needs war as a diversion, he states, as well as a reason to default on debts, issue a central bank digital currency that controls everyone's money, and clamp down on any civil unrest. As economic decline accelerates, EU governments will become more authoritarian, more secretive, and more desperate, and history shows the tendency for government criminal acts to prevent their populations from realizing the depth of the crisis they themselves created. The EU will take whatever steps it can to legislate obedience and silence opposition to maintain control over an unwilling population that sees the government as the problem.

This explains why NATO has been provoking Russia during the Ukraine conflict by attempting to trigger an attack on NATO countries so that war can be declared with Russia appearing as the initiating aggressor. Once war emerges, EU countries would cancel their sovereign debts and fiat currencies, then launch a new system with a central bank digital currency (CBDC) that would prevent capital flight, control citizens' assets, and create a control system for suppressing public behaviors that governments deem undesirable, *such as any criticism of government policies.* After an authoritarian system is set up that keeps the elites in power, the plan is for the cycle to then continue even though Europe is facing "civilization erasure" because of its immigration policies and censorship of free speech.

Through discussions, Armstrong claims that European leaders even believe they will be able to draft their new immigrants to fight this war and actually win militarily against Russia. Why illegal immigrants, who simply came to Europe seeking a free lunch – a better life with all sorts of socialism-paid freebies – would risk their lives for a country they have no allegiance to is beyond anyone's guess. They are not interested in defending Europe so the expectation is unfounded, even bizarrely idiotic. To think Russia militarily weak is also an unfounded notion.

This desperate need to start a war has caused various EU nations to abandon logic and manufacture hysteria by claiming that Russia wants to invade Europe, or in some way take steps towards broader aggression in

Europe. They are getting prepared for a war of their own making. Various parties have been initiating false flag attempts to provoke an actual NATO-Russia conflict. In its long history, however, Russia has never once sought to conquer Europe, but because the EU nations have mismanaged their countries so poorly under EU stewardship, European leaders require an external enemy and must constantly claim that Putin wants to invade.

In October 2025 at the Valdai Discussion Group in Sochi, Russia President Vladimir Putin addressed this hysteria directly, "The ruling elites of united Europe continue to whip up hysteria. It turns out that war with the Russians is practically on the doorstep. They repeat this nonsense, this mantra, over and over again. They can't believe what they're saying, that Russia is going to attack NATO? ... They're either incredibly incompetent if they truly believe it, because it's impossible to believe this nonsense, or they're simply dishonest."

Putin made it clear that the notion of Russia attacking NATO is unfounded, stating that Europe is caught in the grips of irrationality. He understands this as "Rally around the Flag" propaganda, which is necessary to publicly disseminate if the EU and NATO actually want to start a war. Western powers cannot rally their populations to fight unless they are attacked, so they are creating internal unrest and trying to provoke Russia into becoming an aggressor outside Ukraine, but Russia has ignored every provocation designed to trigger an attack in response.

The scheme's proponents also imagine that Russia's $75 trillion in assets – including gold, timber, and oil worth twice the size of the U.S. national debt – will save Europe from financial destruction, preserve the European Union, and bring a renaissance of prosperity through theft, allowing Europe to rise like the Roman Empire so it can rule the world once again. However, we can remember that Reinhart showed that lasting wealth comes from productivity gains, innovation and manufacturing, not from resource extraction or by stealing $300 billion in frozen Russian assets, much of which belongs to individual Russian people rather than the state.

If we look back to 2010, we find U.S. Army announcing that Afghanistan had $1 trillion of untapped mineral deposits, and yet nothing has ever come out of it. Is this not enough evidence? The prospect of a financial bonanza lubricating Europe if Russia is attacked and defeated is a fantasy, just as it is a fantasy that Russia would somehow gain financially from attacking Europe. It has nothing to gain from such a venture since, according to the same reasoning, Europe has little to no assets to seize.

European powers nevertheless try to convince themselves that Russia is an enemy, easy prey, and hope that by conquering it they can partition its territories and seize its assets to rejuvenate their faltering economies, which are burdened by excessive immigration that has harmed national identity, increasing social welfare costs, increasing government interest costs and the inevitability of sovereign debt default. The truth is that they need a war to cancel their debts and stay in power, and so their populations are being methodically marched towards war with propaganda programming. If war occurs, reality will shatter all these EU fantasies.

Viewing matters through a strategic lens, as stated Russia gains nothing from attacking Europe because Europe has no assets to offer. Japan previously attacked Manchuria to obtain resources for its survival, but Russia can extract nothing of value from Europe, which makes the entire invasion narrative a hoax. Putin has pointed out that Russia, spanning eleven time zones, already needs a lot of effort to manage its vast territory, so why would it seek even more? Europe has no resources worth taking.

However, without Russia as an adversary, NATO lacks a mission, military jobs, pensions and weapons expenditures while the EU lacks a distraction from their failed economic policies. Without Russia, there is no one for NATO to fight, thus rendering it purposeless. NATO arguably should have been disbanded in the mid-1990s after the collapse of the Soviet state system, so it is desperately seeking a purpose for its existence when it is no longer needed. Its only purpose is war, and to keep itself being funded paying salaries, pensions and defense contractors it must justify its existence by claiming that Russia wants to invade Europe.

Russia is no longer Communist, no longer poses an ideological threat, has returned to being an Orthodox Christian country, does not subscribe to an ideology of expansion, is not attempting to lead revolutions, and is simply trying to recover from decades of Communist mismanagement. Yet, an irrational hatred against it persists because it previously had a deeply Communist leader (Khruschev) who once said that Russia would bury the West. Russia is not Communist now, and left that fold long ago.

Germany and Japan were both welcomed back into good graces after World War II, but this welcome has not been extended to Russia even though it has been a U.S. ally several times in the past. Armstrong presents documents actually showing that *Russia was previously invited to join NATO*, but conditions have deteriorated to the point where NATO now needs Russia as an enemy for its survival – otherwise it has no purpose and must close up shop. How can you be invited to join NATO and now be considered an enemy so quickly?

The U.S. military-security complex also does not want to lose the highly profitable Soviet enemy. Armstrong says that *NATO even viewed the global warming movement as a threat to its existence* because "climate change" initiatives diverted resources away from military budgets, which threatened NATO. In short, without Russia as an adversary, NATO would be purposeless and on the chopping block.

Even though many smartly recognize that NATO protects its budget, power and even existence by perpetuating Cold War antagonism, the larger and more dangerous self-deception is that NATO would win a war against Russia, and win quickly. The threat is not just Russian nuclear weapons being used, but hyperbaric bombs that can be similarly destructive. Russia has very advanced weapons, and this delusion of a quick win, or win at all, is fueled by Neoconservatives who are blinded by their desire to destroy Russia and ultimately China for various objectives. How many times, in how many situations, have the politicians clamored that a war would be over in days?

Armstrong, having worked with central bankers for over 40 years and been privy to countless conversations with government officials, believes the irrational hatred of Russia is rooted in the fact that the West never militarily conquered Communist Russia. Many Europeans and Americans in elite positions have spent their entire adult lives filled with a hatred of Russians, harboring a loathing for Russia and its leadership that will not depart readily. They are pursuing a vendetta. Hence, there is an American Neoconservative element that will carry this forward and never accept peace with Russia at all. Armstrong suggests some of these Neocons are still seeking an emotional reward for a military conquest of Russia they were denied through peace.

These individuals ignore the fact that the Soviet Union was already defeated when it collapsed, which happened because of financial pressures to match U.S. Star Wars defense spending, and due to covert economic warfare to financially collapse Russia by destroying the value of its ruble. The U.S. made efforts to lower world oil and gold prices (Russia's only two major exports at the time) to reduce its foreign exchange revenues, and to return rubles into Russia in quantity to destroy the value of its currency so as to bankrupt the country. These methods of economic-financial warfare exactly duplicated those used by the ancient Chinese sage Guan Tzu detailed in *The Art of Political Power.*

Which brings us back to the point where we veered off in a sidetrack, which is to explain another type of financial warfare other than efforts to control a foreign nation's interest rates, credit availability, cost of capital and currency fluctuations. This new type of financial warfare, which currently involves a strong naval component, involves China's Belt and Road Initiative.

China's Belt & Road Strategy

What is China's Belt and Road Initiative (BRI)? Launched in 2013 by President Xi Jinping and inspired by the ancient Silk Road, the BRI enhances the connectivity and trade across Asia, Africa, Europe, and beyond through infrastructure, trade, and investments primarily designed to advance China's economic and strategic interests. This connectivity involves an inland component and maritime component (with a naval element) that are designed for China's benefit.

The BRI involves building roads, railways, ports, and energy pipelines, as well as fostering digital and financial integration, across the far eastern region of the Eurasian land mass. It seeks to boost China's economic influence by linking Chinese markets with partner countries, making it easier to sell Chinese goods. Critics argue that it can lead to debt dependency and Chinese geopolitical leverage over these participating countries. Nevertheless, the initiative now includes over 140 nations and has already reshaped global trade routes and investment flows.

When looked at through the lens of geopolitics, the BRI is a development strategy that can gradually shift financial power from the West (U.S./Europe) to China, "equalizing" income in Europe-Asia by draining capital from European economies and redirecting it to Eurasian regions, particularly China.

Let's first examine the inland component of China's BRI. The inland Silk Road Economic Belt focuses on overland routes – high-speed railways, highways, and pipelines – from China's interior through Central Asia

(Kazakhstan, Uzbekistan) to Europe (e.g., Duisburg, Germany). These corridors bypass U.S.-dominated sea lanes, delivering goods 2-3 times faster than maritime shipping with over 17,000 annual China-Europe freight trains (as of recent data), while avoiding naval chokepoints and reducing vulnerability to Western interference and piracy. These inland road and rail pathways shift the economic gravity of trade inland beyond the reach of western navies and air force operations.

The maritime component of the BRI emphasizes sea connectivity, and involves China's "String of Pearls" strategy – a Western analytical term coined in a 2005 U.S. report – referring to a network of ports, airfields, infrastructure projects, military facilities and diplomatic ties that they stretch from China's Hainan Island through the South China Sea and Indian Ocean to the Horn of Africa. To understand the "string of pearls" analogy, think of the sequence of ports located on the perimeter of this region as being pearls on a string. The commercial ports and facilities along this route are essentially dual-use sites that secure energy imports (over 80% of China's oil transits vulnerable routes like the Strait of Malacca), project power beyond the First Island Chain, and counter U.S. and Indian influence. They secure maritime access and chokepoints in the Indo-Pacific, while also protecting China's land-based routes.

The String of Pearls strategy is not an official Chinese doctrine but a Western analytical framework first articulated by Booz Allen Hamilton, a U.S. consulting firm, that was geopolitically analyzing China's efforts to develop this network of commercial and military facilities along key maritime routes in the Indian Ocean Region.

From geopolitical analysis it appears that this strategy aims to (1) secure Chinese trade routes (especially oil imports from the Middle East), (2) project naval power beyond China's "First Island Chain" (Japan-Taiwan-Philippines) and, (3) counter U.S. and Indian influence in the Indo-Pacific. The String of Pearls is a cornerstone of China's integrated grand strategy, blending economic diplomacy with military modernization under President Xi Jinping's vision of a "great rejuvenation of the Chinese nation" by 2049.

China's String of Pearls network aligns with China's broader "far seas" naval ambitions, and emphasizes access to vital sea lines that carry a significant portion of its trade. China has circumvented some oil risk by making pipeline deals with Russia, but these ports will further increase energy security by helping China protect the vital sea lanes that carry oil and gas to China. Basically, the ports help China achieve trade security by insuring an uninterrupted flow of goods along the Maritime Silk Road.

China has one ocean exit – the eastern coast – which is tightly gated by the Philippines, Japan, Taiwan and nearly every Southeast nation, most of which are not friendly to China. Most of China's trade sails within easy striking range of these rival countries. In geopolitics, oceans decide power because they are the roads to trade and wealth, so China has spent billions over the last two decades trying to reduce its risk of having just one sea door by opening up internal trade routes to the Indian Ocean. This is one of the reasons behind the China-Pakistan Economic Corridor project involving the Karakoram Highway through Pakistan connecting to the deepwater Gwadar Port, and rail lines (and pipelines) across Myanmar's mountains to reach its western coast.

China has formulated its own naval strategy called A2/AD to complement the String of Pearls, and because China is now our manufacturing and financial rival this bears more alarm than any focus on Russia. China's A2/AD or "Anti-Access/Area Denial" naval strategy deters U.S. intervention within China's First Island Chain (Japan-Taiwan-Philippines) by preventing entry (Anti-Access), and restricts American or other foreign military operations (Area Denial) via Chinese missiles, submarines, and island bases. The design of its strategy is *sea denial* inside the First Island Chain and *sea control* inside the nine-dash line to make intervention so costly inside "Chinese waters" that America cannot politically protect its Allies (Japan, Taiwan, Philippines). Viewed by China as defensive "active defense," it is seen by the U.S. as offensive, challenging Mahan's sea-control doctrine that has underpinned American naval dominance for generations. For China, this is rightfully establishing its sphere of influence just as the U.S. protects its interests in the Western hemisphere (its neighborhood) as outlined in the Monroe Doctrine.

The Anti-Access (A2) aspect of this strategy creates a "keep out" zone meant to prevent adversaries (e.g., U.S. Navy) from entering China's regional waters or airspace. The Area Denial (AD) aspect of this strategy prevents an opposing force from operating freely within this area even if they manage to enter it. China's methods to limit an adversary's operational effectiveness within the First Island Chain (Japan-Taiwan-Philippines) makes it extremely costly and dangerous for any enemy to maneuver and project power within the contested zone. Essentially, China's A2/AD strategy is about creating a heavily defended "bubble" or "kill zone" where an adversary's military advantages (like powerful aircraft carriers) are neutralized. Toshi Yoshihara and James Holmes, in their 2010 work *Red Star Over the Pacific*, identified the Chinese A2/AD strategy as systematically pushing U.S. forces from the Western Pacific.

Overall, while China's economic moves build regional and global influence, A2/AD is the world's premier anti-access system that advances China's "active defense" doctrine to protect its territorial claims and ensure sovereignty over core interests like the South China Sea. Every nation tries to do this.

Naturally, China hopes to project naval power and influence beyond the First Island Chain, so A2/AD can also be viewed as a foundation for eventually projecting power beyond the Second Island Chain (Guam). In this way, China's navy can transition from simply a coastal force to a blue-water force that can operate far from home waters (such as conducting anti-piracy missions in the Gulf of Aden).

Since China's ports can also be used as forward bases for naval operations and surveillance, India and the U.S. both view China's String of Pearls network as a potential military threat. While China sees A2/AD as defensive, protecting its sovereignty, the U.S. perceives A2/AD as offensive, aimed at regional hegemony. A2/AD jeopardizes U.S. carrier strike groups and bases (e.g., Guam, Okinawa), with missiles potentially outranging U.S. assets. In short, China's A2/AD seeks to secure regional control and deter U.S. intervention through asymmetric methods, while the U.S. sees it as a threat to its own global naval primacy, prompting a shift toward coalition-based deterrence and American countermeasures through technology.

The U.S. Navy's aircraft carriers and 300+ ships dominate open-ocean warfare, with freedom of navigation operations enforcing a rules-based order. China's String of Pearls disrupts this by introducing a multipolar maritime environment, and its A2/AD is a direct challenge to U.S. naval dominance and the post-WWII rules-based order that relies on freedom of navigation and forward-deployed forces. As a new challenge, the U.S. must adapt to China's moves or lose its Pacific dominance to China.

Viewing it strategically, China's String of Pearls may perhaps be the foundation of a 21st century maritime empire just as Britain, Portugal and Spain created in the past. Its moves challenge the old paradigm of U.S.

supremacy, which replaced British naval supremacy. British supremacy replaced the earlier historical rivalries of the Dutch, Portuguese, Spanish, and British all vying for sea power to secure trade routes as roads to prosperity. In particular, it mirrors Britain's 19th-century naval strategy (coaling stations, trade dominance) but offers China much more leverage.

The pearl ports provide logistical hubs for the People's Liberation Army Navy, extending its ocean capabilities because the pearls are effectively dual-use facilities that allow gradual militarization under commercial guises – enhancing deterrence without overt aggression. While these pearls signal China's shift from merely coastal defense to a true blue-water navy (global naval presence), it would be a great leap to say that the mission of China's navy has switched from the support of coastal defense and land-based combat to engagement in decisive ocean battles striving for command of the seas. Its moves are far from mimicking the British Royal Navy's *"Rule Britannia"* command of the sea doctrine or the *"Pax Americana"* strategy where the U.S. Navy has acted as global policeman.

Geopolitically, the BRI functions as financial warfare, shifting capital flows and financial power from West to East: China exports low-cost goods via BRI land and sea routes while being paid in foreign currency and attracting Western investment, gradually raising its average national income and GDP per capita at Europe's relative expense. By continually shipping goods eastward while becoming a necessary supplier who is paid higher prices over time, China is engaged in an equalizing arbitrage that will shift its economic statistics upwards while also lowering those of Europe. The flow of funds it produces mirrors historical trade imbalances like Britain's 19th-century silver drain to China to buy tea. Furthermore, the new trade routes will draw Europe and Asia closer together in a way that economically increases their mutual power and prosperity while leaving the United States on the outside looking in.

The BRI trade routes will over time essentially equalize Eastern and Western economic development by weakening Western financial dominance and boosting Eurasian economies. Over time they will "equalize" income and GDP on both sides by draining capital from Western economies and redirecting it to China-aligned regions that supply goods. One side goes up, the other side goes down as the two sides create stronger linkages that equalize them. China will become richer, and relatively speaking, western nations in Europe will become less financially viable, especially if they lose manufacturing advantages due to facing lower costs from Eastern competitors.

By making the transportation of its goods to western customers easier and cheaper, China will see an increasing demand for its manufactured goods, and then experience the wealth accumulation effect identified by Reinhart. Hence, the BRI "equalizes" income asymmetrically with China gaining more (expands influence, secures resources) in the overall system.

Some BRI partners (e.g., Pakistan, Ethiopia) will see short-term growth from the BRI but risk future debt crises in owing funds to China for their infrastructure projects. Europe is divided: Some regions will benefit (e.g., the Piraeus Port in Greece) by gaining investment but risking long-term dependency, while others will lose manufacturing jobs to China. The BRI may weaken the West's financial stranglehold on Eurasia, and will shift the wealth to whomever controls capital.

The Belt and Road Initiative also accelerates de-dollarization by offering yuan-denominated infrastructure loans to foreign countries along its routes, reducing reliance on the USD and Western financial systems like SWIFT that control global finance. The BRI, with both land and sea routes that bypass western control, should therefore be seen as part of a larger financial struggle over global capital flows, and essentially a key to the wealth and future of nations just as maritime disputes were in the past.

De-Dollarization

When analyzing China's strategic challenge, most commentators overlook a fundamental truth: the post-World War II economic order – Bretton Woods (1944) and later the petrodollar system (1970s) – created an implicit bargain. The United States provided global public goods – security, open sea lanes, suppression of piracy, and freedom of navigation through key chokepoints (Hormuz, Malacca) – in exchange for the dollar's status as the world's primary reserve currency.

The U.S. basically became "Policeman of the Seas" in a deal that benefitted everyone. This arrangement produced a "peace dividend" for all the nations of the world, at U.S. expense, which allowed other nations to develop economically without maintaining large navies of their own, while the United States gained the "exorbitant privilege" of financing deficits by issuing the currency everyone needed.

Because countries could trust that ocean freight would reach them, they could cut costs by stretching their supply chains to include overseas suppliers, which produced the globalization trend where countries shipped their manufacturing offshore. The so-called "free trade" trumpeted by globalization advocates aiming to make a profit ended up hollowing out the American middle class and industrial base on which American economic and military power depends, and the policy of U.S. firms offshoring American manufacturing jobs built the Chinese economy. In other words, the U.S. created its own competitor due to short-term interests rather than wise strategy. It now needs to rebuild its manufacturing base.

Today, if other nations no longer want to use the dollar for global trade such as petroleum purchases, the U.S. would end up spending billions to protect international economies without any return, and then be policing the seas for free. In reviewing history, we have clearly seen that the design and usage of a navy is overwhelmingly due to fiscal constraints and returns. When the economic returns of a navy decline, it inevitably shrinks in size. To analyze this situation we cannot remain locked solely in the barrier of the military domain.

The petrodollar system emerged after President Nixon ended dollar-gold convertibility in 1971. In 1973–74, the U.S. reached an agreement with Saudi Arabia (later extended to OPEC) that oil would be priced and traded exclusively in dollars. In return, Washington provided military protection, arms sales, and geopolitical support to Gulf producers. This created permanent structural demand for American dollars: every oil-importing nation had to accumulate USD reserves (mostly U.S. Treasury securities) to pay for energy. Petrodollar recycling – oil exporters reinvesting surplus dollars in U.S. assets – helped finance American deficits and kept American interest rates low, effectively subsidizing the massive military (especially naval) posture required to protect the system itself.

That system is now fracturing, largely due to self-inflicted American wounds. By weaponizing the dollar and the SWIFT messaging system through sanctions – most dramatically against Russia after 2014 and especially 2022 – Washington convinced much of the world that holding dollars and Treasuries carries political risk. Nations concluded they could be cut off or have their assets frozen simply on Washington's whim. The result has been rapid de-dollarization:

- Russia now sells oil and gas in non-dollar alternatives such as rubles, yuan, Euro and rupees, thus bypassing the dollar and SWIFT system entirely.
- Russia and China are using direct currency exchanges for trade, as well as other nations. Dozens of bilateral trade agreements (India–UAE, China–Brazil, Russia–India, etc.) have eliminated the dollar entirely.
- China prices some oil contracts on the Shanghai exchange in yuan, convertible to gold.
- China's Belt and Road initiative offers infrastructure loans in yuan, reducing dollar dependency in dozens of countries across Eurasia and Africa.
- Central banks are now buying gold at the fastest pace since records began and reducing Treasury holdings.

These and other developments have accelerated dollar avoidance. The reason many nations started bypassing using the dollar was because of U.S. moves to weaponize its currency and SWIFT. Many countries thereafter started looking for USD alternatives so that they could not be blackmailed by having their assets frozen or taken. They saw the sanctions against Russia and reasoned, "This might happen to us, too, so let's lower the risks by trading in alternative currencies and parking some money in gold."

The USD is therefore coming under attack from self-inflicted wounds. China's naval plans in the South China Sea challenge the U.S. dominance in the region, too, and if those trade routes become contested the dollar's security backing further weakens, which is yet another threat to the dollar. Confidence in the dollar's reliability is waning. More and more foreign investors are divesting from U.S. Treasuries in favor of gold as a banking asset, yet another source of deteriorating demand for dollars.

Under Basel III rules (implemented 2019), gold is now a Tier-1 asset counted at 100% of market value for central bank core capital requirements – treated essentially as cash – while U.S. Treasuries have lost their automatic risk-free status for many investors. Other nations are now investing in gold instead of U.S. Treasury bonds in order to diversify their risk exposure, so de-dollarizing is proceeding due to a large number of reasons.

The BRICS nations now aligning are holding more yuan, euros, and gold – and less dollar assets – softening the dollar's role as the top global reserve currency. The dollar's share of global reserves has fallen from ~70% in 2000 to ~58% today and continues declining.

Without the captive demand once provided by petrodollars, the United States can no longer finance trillion-dollar deficits indefinitely by printing money. Interest payments on the national debt already consume more than the entire defense budget, and foreign purchasers are retreating. The "peace dividend" the U.S. Navy once provided to the world – guaranteeing everyone's trade while others free-ride – is becoming unaffordable.

This has direct implications for naval strategy.

First, the fiscal foundation of American sea power is eroding. A navy built on the assumption of endless deficit spending must shrink or transform because the monies will not be there. The U.S. must reshore its manufacturing, bringing it back home to rebuild the country and fiscally fill coffers, which will require an overhaul of the U.S. tax system to bring businesses back from lower-cost countries. Tariffs can fund an American economic renaissance.

Furthermore, the U.S. must also once again become an innovator nation that invents new products and solves difficult problems, rather than handing this title over to other nations. An innovative engineering mindset will be required to accomplish this redirection, thus requiring universities to put more emphasis on R&D and engineering/science educations (and innovation methods such as TRIZ).

The U.S. would benefit from restoring a national emphasis on R&D (e.g., Bell Labs) along with small business creation since autonomous entrepreneurships accounts for the majority of job creation in a country. The trend must turn to production and away from financialization paper games and financial services as a large source of national income, and America must undertake moves to correct a misdirected educational system that does not feed the sources of national prosperity.

Second, the strategic rationale for various naval functions is weakening. If global trade is no longer conducted predominantly in dollars and if key routes are contested by a peer competitor (China), why should American taxpayers continue paying to police sea lanes where this primarily benefits others? The U.S. practice of patrolling the world's oceans as a policeman to protect all trade, thus providing a peace dividend to all nations, no longer benefits the U.S. as it once did. It is no longer *affordable*.

With multiple currencies now being used in world trade, and several countries with large militaries able to challenge the U.S., a unipolar world with a single supreme hegemon is now no longer possible. The U.S. must therefore re-think its grand strategy, including the role of the navy in relation to any new currency backing by world nations. Another mission is that America's financial system become aligned with national purpose.

The rejiggering of world power will be tied to currencies being backed by gold and silver in the future. If the dollar loses its importance in reserve status and a multipolar currency system emerges (whether yuan-based, gold-linked, or basket-based), the United States will need to secure access to strategic commodities – especially gold and silver – with far fewer financial resources. Therefore the U.S. must examine its relationships with all nations that mine and export precious metals because those relationships will take on new importance. If gold is to become the basis of a new monetary system, U.S. strategy must recognize that African gold producers will come under incredible take-over pressure through various foreign mechanisms.

The navy must analyze the risks to ocean-front nations that export these metals, such as Mexico which is a large silver producer, examining the implications for the navy's role in protecting their ports and trade routes. An example: Mexico, Peru, Ecuador, Argentina and Chile together account for nearly 60% of world silver production. While Mexican ports do not belong to China's "string of pearls" and China does not own Mexican ports, Chinese companies are key players and long-term operators within strategic Mexican ports, and thus can affect silver trade. The U.S. must definitely secure its sources of strategic and monetary minerals because outsourcing its needs (of things that protect the people or sustain our way of life) to untrustworthy countries can resulted in being blackmailed or suicided. Critical minerals are the new oil (later this may become water), and the U.S. must secure its supply chain, especially when it considers the possible need to fight wars.

Third, the technological and tactical environment has changed irreversibly. Missiles, hypersonics, drone swarms, and autonomous systems have made large surface combatants – especially aircraft carriers –

exquisitely vulnerable. Houthi attacks in the Red Sea (2023–2025) using \$20,000 – \$2 million weapons forced shipping to reroute around Africa, proving that cheap, numerous, precise systems can deny sea control to even the most advanced navies. Iranian and Chinese anti-ship ballistic missiles, combined with satellite targeting, push carriers outside their effective range or turn them into multi-billion-dollar targets.

Missiles and drones are now decisive in air and naval warfare, which previously focused primarily on tonnage. The U.S. is still too focused on carrier thinking, but cannot stay wed to anachronistic WWII naval strategies and structure. The future is shifting from carrier-based and submarine-based warfare to something far more mobile and far less expensive.

Naval supremacy is becoming dependent on networked warfare that integrates sensors, satellites, unmanned systems, and AI to achieve an information dominance for warfare dominance. Success hinges on information, swift decisions, appropriate actions, and adaptability – cutting-edge technology and nimble strategy. The use of AI in naval warfare enables autonomous swarms, which can attack by air or sea, capable of overwhelming traditional defenses through their coordination and sheer numbers. These systems operate at speeds that far exceed human decision-making capabilities, and thus require new command structures and rules of engagement that can take advantage of our own machine-speed decision-making.

Cybersecurity must also become a core competency, with hardened systems on ships designed to operate in contested electromagnetic environments, and officers trained for scenarios where traditional communications are jammed or compromised.

Drones, autonomous vehicles, robotics, missiles and space-based energy weapons will play critical roles in the future of naval warfare, and while aircraft carriers will still retain many functions they are increasingly susceptible to attack. Since other countries now have same technological capabilities as the U.S. in many of these weapons areas, and counter-measures to the U.S. military are now low-budget weapons that can be easily produced in mass quantities, which country will not develop them and use them? An entire new set of military strategies need to be devised.

Fourth, the rising power is clearly China rather than Russia. China's rapid naval expansion – three carriers in service, a fourth under construction, Type 055 cruisers, a doubling of the size its Marine Corps, and a growing nuclear submarine force – occurs against this backdrop. Beijing can now contest the Western Pacific and is extending its reach through the Belt and Road Initiative and port investments from Pakistan to Djibouti to Cambodia – and increasingly in Latin America (where Chinese firms operate key terminals in Mexico, Peru, and elsewhere). The U.S. must determine new strategies in relation to these developments, while understanding that China's national prosperity was fostered by our own actions while its economy is deeply linked with ours.

Krepinevich and Tangredi (*Maritime Strategy in a New Era*, 2012), as well as others, have warned of "missile salvo warfare" where victory goes to the side that can fire first and saturate enemy defenses. The old Mahanian focus on concentrated battle fleets and decisive surface engagements is increasingly obsolete. Navies must adapt to the low-cost asymmetric threats – drones, cyber-attacks, and hypersonic weapons – that level the playing field between great powers and smaller adversaries. One $20K-$2M drone or missile can mission-kill or sink a $10-15B carrier or destroyer. Future naval power will likely be distributed across hundreds or thousands of smaller, unmanned, or lightly manned platforms – air, surface, and subsurface – networked by AI and space-based sensors. A new fighting paradigm demands development because the delusion of building multiple carriers to win in this environment can quickly result in bankruptcy.

The U.S. Navy's own concepts – Distributed Lethality, Ghost Fleet, large unmanned vessels – acknowledge this reality. Carriers will not disappear, but they will in all probability become floating command centers and drone motherships rather than the primary strike arm. Long-range missiles and unmanned aircraft launched from dispersed, cheaper platforms will do much of the fighting.

Long-range strike capabilities, utilizing missiles and extended-range drones, will complement and sometimes replace short-range carrier aircraft, enabling strikes from beyond A2/AD zones. The Navy's current ship goal includes over one-hundred large unmanned vessels, which signals a realistic recognition that future fleets will contain hybrid human-machine systems. The potential of unmanned air and undersea drones cannot be understated in any new fleet configuration, but the ease of producing such systems (e.g., drones) questions whether the U.S. Navy can retain a significant technological edge or become defeated by sheer numbers in an attack.

Strategic Implications and Required Adaptations

1. Fiscal reality forces prioritization. Traditional sea control is becoming prohibitively expensive, particularly as the Petrodollar status declines and the U.S. loses its financial hegemony. The United States must live within its means, cut non-essential spending, and focus the defense budget on capabilities that provide asymmetric advantage (unmanned systems, hypersonics, undersea warfare, space/cyber). The U.S. Navy has played a vital role not just in maintaining the global economic order, but in producing a valuable global peace dividend by enforcing the "Pax Americana," patrolling the South China Sea, combating Somali piracy, and ensuring safe passage of oil tankers while global oil traded in U.S. dollars. Petrodollar recycling (where oil profits are reinvested in U.S. Treasuries) kept American liquidity high, but this seems to be ending.
2. With limits on military spending, it becomes obvious that entering into a British style "balance of power forever conflict" where you never actually solve a conflict but manage it (wars not meant to be won, but be continuous as "managed conflicts"), is a sure recipe for national bankruptcy or ending a nation. Peace is no longer just a preference, but a necessity preferred to permanently managed conflict zones and their chaos. For the ethnically, racially and religiously diverse United States to regain its own internal peace and prosperity where people of vastly different backgrounds are able to live together and coexist peacefully side by side, it must eliminate both foreign and *domestic* influences constantly working to fragment national cohesion. Other nations commonly pit countries or various forces against each other in perpetual war so that no one is ever strong enough, united enough or peaceful enough to break free of backwardness (to achieve prosperity) or foreign looting, so the U.S. must oppose any internal unbalancing strategies used against it that work to limit American peace and prosperity. This includes anti-Whiteism, anti-Christianity, non-merit based favoritism and the slow but definite, gradual infiltration of personnel and ideologies aimed at controlling U.S. foreign policy and weakening the country by fracturing its peace, economic prosperity,

internal social cooperation or its advancement. Sovereignty, for instance, is your borders. We must recognize that there are always geopolitical playbooks running against us – extensively planned long-term subversion operations to weaken us, fragment us, or exploit us. There are on-going foreign interventions in rigging American elections, attempts to corral and then direct Gen Z and other youth segments as shock forces and ramming rods to overthrow what foreign powers dislike, and even operations like Russiagate where foreign intelligence agencies clearly cooperated in trying to topple a U.S. President. There are definite gradual attempts at ideological subversion to prevent the U.S. from becoming stronger, and targeted infiltration strategies where harmful players are gradually placed into specific positions of power and influence, or native leaders are selectively targeted and then compromised. America is a big prize to certain global players if they can conquer or subordinate even just a small piece of it through some fashion, and the military's job is to stay on alert to oppose this by piercing through the messaging of corrupt media or psyops designed to mislead the public and direct its anger through false narratives and ideological capture. Such antagonistic forces must be neutralized; some must be dismantled from within the U.S. itself.

3. Grand strategy must shift from global policeman to offshore balancing, relying more on allies (AUKUS, Quad, NATO maritime components) to share burdens in their own regions. They must finance their own militaries and be willing to fight themselves rather than the U.S. taking the lead and obligation. It currently funds 60% of NATO expenditures but does not need it for anything. The focus of U.S. military strategy should switch to the Western hemisphere, America's sphere of interest, rather than "exporting democracy," color revolutions for regime change, and intervening to fight wars for others at America's expense. Intervention in "forever conflicts" at someone's behest is neither profitable for the U.S. nor fiscally sustainable, and the history herein already reveals what happens to navies when governments can no longer financially sustain them.
4. Technological adaptation is non-negotiable. The future fleet will be hybrid: fewer supercarriers, more missile-armed frigates, large unmanned vessels, drone swarms, and next-generation submarines. Cybersecurity and resilience in contested electromagnetic environments

are now core warfighting competencies. This is now the Age of the Drone, which fundamentally changes military doctrine, so I have detailed one unique electro-magnetic vortex method for dealing with drones in *The General's Playbook*. Ships are now in danger not just from missiles and directed energy weapons but from drone operators situated miles away in safety. Lasers and electromagnetic assaults (electronic warfare) and directed energy weapons are simple protective mechanisms against drones and some types of missiles.

5. Economic and education policy are now national security imperatives. Reshoring manufacturing, rebuilding the manufacturing base, reforming taxation policies to favor domestic production, and reorienting upper education toward engineering, entrepreneurship, and STEM are as important as ship counts. The manufacturing base is a pool of practical capability and problem solving that should not be lost; as Alexander Hamilton said, the wealth of America is its productive power of labor to build, innovate, invent and produce. It was an act of insanity to destroy this base by shipping it away, and to switch to an economy based on financialization (see *Bankism*) rather than manufacturing and technological-industrial innovation. Now we must totally rebuild our economy. Across the world, the normal competitive angle of seeking low-cost labor is being replaced by competition for technological dominance and innovation waves, and we need to be a the forefront. Deindustrialization is economic suicide – a nation's own workers should come first, and an economy should favor working people and their human dignity. Economic sovereignty and resilience also requires control of basic raw materials necessary for national survival, such as copper and silver, and the maintenance of certain manufacturing capabilities that protect national security through strategic autonomy. No nation should be dependent on other countries for things fundamental to its sovereignty, including its growth.
6. Sea control remains linked to trade dominance, but the critical arteries are no longer just oil tankers – they are undersea internet cables carrying 95% of global data. Protecting these cables may become the 21st-century equivalent of protecting trade routes, while severing them is a form of economic warfare equivalent to blockading an enemy's commerce.

The era of unipolar American sea power sustained by dollar hegemony is ending – not primarily because of Chinese shipbuilding, but because the economic foundation that made 600-ship navies and global policing affordable is collapsing under its own weight and strategic missteps. However, the U.S. Navy can remain the world's most powerful if it ruthlessly modernizes, distributes lethality, embraces unmanned systems, and aligns its strategy with a multipolar, de-dollarized world.

Failure to do so risks strategic irrelevance – or worse, defeat by a swarm of cheap drones launched by an adversary who learned the lessons we taught the world. The traditional naval battles at sea that our brilliant commanders have fought so valiantly will transform into a new form as they have always done over time. We must therefore realistically recognize that the ships and weapons which once made naval dominance possible are now less effective. Force can now be distributed across hundreds of unmanned platforms rather than concentrated in single hulls, so we must take appropriate steps to prepare for this future to the right extent and in the right way.

ABOUT THE AUTHOR

An ex-Wall Street analyst who has studied world supremacy cycles, geopolitics, and military affairs, Bill Bodri is the author of several highly acclaimed non-fiction books:

The General's Playbook: The Signature Moves of History's 15 Greatest Generals – This strategy book distills the decisive, repeatable "signature moves" of military history's most successful commanders, such as Alexander's oblique hammer-and-anvil technique, Hannibal's double envelopment move, and the commonly used winning tactics of Caesar, Napoleon and others. Each general used multiple tactics again and again that are clearly identified and explained. By stripping each general's campaigns to their most commonly used and transferable ideas, the book functions as both a concise history of military operational art and a practical playbook for anyone who must out-think and out-maneuver an adversary on the battlefield. An insightful last chapter on electronic warfare against drones, weather engineering and other rare topics.

The I Ching Revealed: Tap Into the Five Secret Patterns Underlying the 64 Hexagrams – This groundbreaking translation of the I Ching (*Yijing*), based on the oldest version found in the Mawangdui manuscript, reveals it to be a collection of sixty-four essays on topics of interest that progress in a particular pattern that has been decoded rather than a divination manual. Many hexagrams contain commentary on military topics because it focuses on the Zhou dynasty's conquest of the corrupt Shang dynasty of China. Hexagrams 5, 7, 8, 10, 11, 13, 15, 24, 27, 34, 35, 41, 42, 43, 58, 63 and 64 all concern military affairs.

The Art of Political Power: A Translation and Commentary of The Book of Master Guan (The Guan Tzu) – A translation of an important Chinese strategic classic, often considered superior to the *Sun Tzu* because of its higher-level perspective on geo-political moves and political strategies rather than just military tactics. Everyone reads *Sun Tzu* but they should also read *Guan Tzu (Guanzi)* for an elite geopolitical understanding of national economic, political and military strategies. It's about structural power theory. The *Guan Tzu* reveals *economic warfare principles* for conquering countries without fighting, and provides lessons for how ordinary leaders can make themselves into dominant political leaders, as well as how to transform a country into a superpower. An essential read for anyone interested in geopolitics, economics, military strategy, political power, and the hidden forces that shape world events. These lessons are a companion to Sun Tzu's writing on military tactics, but view matters from the higher ruler's perspective.

Culture, Country, City, Company, Person, Purpose, Passion World: The Grand Strategies and Unifying Principles Behind the Groups Which Rise and Thrive – Why do civilizations and countries rise and fall? What principles determine which companies thrive while others collapse? This historical analysis reveals the underlying patterned principles of growth and prosperity ruling various organizational structures (civilizations, empires, countries, cities, companies), and how specific economic cycles which affect these structures can be harnessed by individuals, companies or countries as forces for generational wealth. It contains much of the information people want regarding civilizational and national collapse patterns. Military campaigns can win an empire, but what do you do afterwards? This book tells you, and reveals why we are seeing nations abandoning common sense for chaos and suffering monetary, immigration and defense problems at a time when several cyclical change patterns coincide. Its contents guides those who hold Themistocles's objective: "I never learned how to tune a harp, or play upon a lute, but I know how to raise a small and inconsiderable city to glory and greatness."

Neijia Yoga: Nei Gong for Yoga and the Martial Arts – The martial arts emphasize ideal body alignments for physical movements designed to express tremendous muscle power and energy for self-defense and offense. This book will guide you to reach the highest levels of martial arts using internal energy exercises that move your internal energy in tune with your movements. You will learn how to amplify Qi through bone, muscles and organs for striking or protection. Within are the special Chinese esoteric techniques for cultivating the Qi (Prana) of your energy meridians that are Qi circulatory pathways; muscle force transmission lines; *bindus, marma* or acupuncture points; appendages such as arms and legs; body cavities and simple body parts (such as the ears, eyes, teeth, penis, and so forth).

Super Investing: 5 Proven Methods for Beating the Market and Retiring Rich – An investment guide revealing five time-tested techniques that have consistently outperformed markets over the past century, an amazing feat. These aren't fads – they are enduring wealth models. The book provides methods that have achieved consistent 20-25% annual compounding rates across decades.

High Yield Investments, Hard Assets and Asset Protection Strategies – This book teaches you how to invest in different types of high-yield investments (many you've never heard of but are easy to buy) – along with strategies for protecting these investments from institutional risks and government monetary changes through hard assets. The major focus, besides hard asset safety, is on creating a bulletproof portfolio of high-yield income that is essentially a cash-compounding machine for consistent passive income.

Bankism: How the Government's Bank-First Policies are Destroying the Nation and How to Survive the Aftermath of a Coming Dollar Collapse – A reset is coming to the world's financial system. Part of the solution will involve the self-funding state bank, such as the Bank of North Dakota, and precious metals. The safest course of action for the transitional period ahead is to own physical gold and silver, as well as select cryptocurrencies (perhaps XRP), in order to safely pass through the coming collapse of the dollar, inevitable decline in asset prices, and destruction of the U.S. economy as various non-dollar payment systems go online that will supplant the dollar and reduce its predominance in world trade. Here is a history of what brought us to this state of affairs and what to do to move ahead.

Sport Visualization for the Elite Athlete – The secret mental edge used by Olympic champions and professional superstars is visualization practice. This performance-enhancing guide reveals how to harness visualization techniques and mental rehearsals to dramatically improve athletic abilities – whether in competitive sports, martial arts, yoga, or other physical disciplines. This is one of the psychological components that separates world-class performers from mediocre players. Elite athletes use it because visualization works in improving performance.

Quick, Fast, Done: Simple Time Management Secrets From Some of History's Greatest Leaders – This short but remarkable guide on productivity hacks and time management methods offers techniques for accomplishing more in less time so that you will improve your productivity. Time is the battlefield, so stop drowning in tasks and reclaim your life. Learn to get more done in less time, be more productive, and put your life back on track while others remain trapped in cycles of endless busyness and burnout.

Move Forward: Powerful Strategies for Creating Better Outcomes in Life – This isn't about goals. It's about transformation. Inside are proven strategies for overcoming the flaws of your genetic inheritance, breaking ingrained habits, and changing a "fated astrological fortune" to create radical personal improvement in areas such as health, business, habits, talent development and spiritual growth. These techniques will help you rewrite your life's script by overcoming limitations. You don't need permission. You need strategy and tactics to achieve better life outcomes.

www.ingramcontent.com/pod-product-compliance
Lightning Source LLC
LaVergne TN
LVHW022211200926
848481LV00006B/1920
* 9 7 8 1 7 3 7 0 3 2 0 8 3 *